Advances in Quantum Transport in Low Dimensional Systems (AQT2017)

Journal of Physics: Conference Series Volume 964

London, United Kingdom
4 – 5 September 2017

ISBN: 978-1-5108-5931-9
ISSN: 1742-6588

Printed from e-media with permission by:

Curran Associates, Inc.
57 Morehouse Lane
Red Hook, NY 12571

Some format issues inherent in the e-media version may also appear in this print version.

Printed by Curran Associates, Inc. (2018)

For permission requests, please contact the Institute of Physics
at the address below.

Institute of Physics
Dirac House, Temple Back
Bristol BS1 6BE UK

Phone: 44 1 17 929 7481
Fax: 44 1 17 920 0979

techtracking@iop.org

Additional copies of this publication are available from:

Curran Associates, Inc.
57 Morehouse Lane
Red Hook, NY 12571 USA
Phone: 845-758-0400
Fax: 845-758-2633
Email: curran@proceedings.com
Web: www.proceedings.com

Table of contents

Volume 964

Advances in Quantum Transport in Low Dimensional Systems
4–5 September 2017, London, UK

Accepted papers received: 29 January 2018
Published online: 2 March 2018

Preface

Papers

IOP Conf. Series: Journal of Physics: Conf. Series **964** (2018) 011001 doi:10.1088/1742-6596/964/1/011001

Preface

The conference titled, "Advances in Quantum Transport in Low Dimensional Systems (AQT2017)" was held at University College London, during September 4-5, 2017. The conference was organised by the Institute of Physics (IOP) Nanoscale Physics and Technology Group, and was sponsored by the Oxford Instruments (primary sponsor), National Instruments, Nanomagnetics Instruments, Zurich Instruments, University College London, Specs, IOP Nanoscale Physics and Technology Group and IOP Quantum Electronics and Photonics Group.

The idea of organising a conference dedicated to advances in the transport in low dimensional systems came from the fact that advances in fabrication of solid state systems have reached the limit in which quantum effects cannot be ignored anymore. This has opened new perspectives for the development of new fundamental physics and products for commercial applications. Due to the rapid development of quantum technologies active collaboration between theory, experiment and industry is essential to meet the future demands. In solid state quantum technologies, there are a number of challenges, at the level of both theory and experiments, which need to be discussed openly in a platform shared by theorists, experimentalists and industry people.

The aim of this conference was to bring together experts from various areas of solid state quantum physics with specialisations in quantum transport in low-dimensional systems (2D, 1D and 0D). The conference was well attended especially research students actively participated. AQT2017 provided an opportunity for both academia and industries to seek new future possibilities in the emerging solid state quantum technologies and nano devices with special interest in the spin physics of low-dimensional systems. The conference provided great opportunity for enriching the knowledge of young researchers and students through interaction with the experts in concerned areas.

We covered the following topics during the conference: transport in 1D, 2D and 0D systems, mesoscopic effects, quantum pumping, quantum and spin Hall effect, spin or charge entanglement and topological insulators and superconductors.

We were privileged to hear talks from the experts in the field. We are grateful to all the invited speakers for their exciting, enriching talks and sharing their current research activities with us. In this regard, we thank Prof Sir Michael Pepper, Prof Jainendra Jain, Prof Moty Heiblum, Prof Seigo Tarucha, Prof Irfan Siddiqi, Dr Masaya Kataoka, Prof Karl-Fredrik Berggren, Dr Konstantin Matveev, Prof Charles Smith, Prof Chris Ford and Prof Irina Grigorieva for their enlightening presentations on respective research fields. We also thank Dr Graham Batey, Oxford Instruments for a fascinating talk on the history of dilution refrigerators. In addition to the technical sessions comprising of presentations by the experts and contributed speakers, a poster session was organised as a forum for networking as well as opportunity to interact with the experts in the field.

We are grateful to the Institute of Physics for all the help in organising this conference, and particularly agreeing to publish the proceedings of this conference in Journal of Physics: Conference Series, IOP Publishing.

We thank members of the advisory and local organising committee, attendees and all the sponsors without their active participation and support, organising this conference would not have been possible.

Sanjeev Kumar
Chair, Advances in Quantum Transport in Low Dimensional Systems
September 4-5, 2017, University College London, London, UK

IOP Conf. Series: Journal of Physics: Conf. Series **964** (2018) 011002 doi:10.1088/1742-6596/964/1/011002

Peer review statement

All papers published in this volume of *Journal of Physics: Conference Series* have been peer reviewed through processes administered by the proceedings Editors. Reviews were conducted by expert referees to the professional and scientific standards expected of a proceedings journal published by IOP Publishing.

IOP Conf. Series: Journal of Physics: Conf. Series **964** (2018) 012001 doi:10.1088/1742-6596/964/1/012001

Role of random edge-disorder on the transport properties of ultra-thin zig-zag graphene nanoribbons

J P C Baldwin[1] **and Y Hancock**[1,2]

[1]Department of Physics, University of York, Heslington, York YO10 5DD, United Kingdom
[2]York Centre for Complex Systems Analysis, University of York, Heslington, York YO10 5GE, United Kingdom

E-mail: `y.hancock@york.ac.uk`

Abstract. We report on an introductory study used to gauge the significance of random weak-edge disorder on the coherent transport properties of ultra-thin zig-zag nanoribbons (ZGNRs) beyond the simple (*i.e.*, first nearest-neighbour) tight-binding approximation. Such extensions include up to third nearest-neighbour hopping in an extended tight-binding model, as well as a mean-field Hubbard-U. The effect of the random weak-edge disorder causes charge-carrier localization that reduces the conductance about the Fermi energy in all of the systems studied. In the non-interacting systems, the extended tight-binding model is found to be more robust against disorder due to the increased kinetic degrees of freedom. Localization effects from the random weak-edge disorder are found to compete with the mean-field Hubbard-U resulting in spin-dependent conductance properties.

1. Introduction

Edge roughness [1] and chemical disorder due to highly reactive edges [2] are common issues in the fabrication of graphene nanoribbons requiring realistic modelling and fast simulation for future device design. Within this context, the role of disorder on the transport properties of low-dimensional quantum systems becomes a key question, and thus, in this paper, we explore its effect in ultra-thin ZGNRs. Ideal ZGNRs are predicted to have antiferromagnetic order across the ribbon-width commensurate with Lieb's theorem for a half-filled bipartite lattice, and a band gap that varies inversely as a function of the increasing ribbon width [3]. With experimental measurements also showing magnetic signatures in graphene nanoribbons with ZGNR edges [4], the interplay between magnetic effects, system-dimension and random edge disorder becomes pertinent and timely considerations.

There are few density functional theory (DFT) studies on disordered ZGNRs most likely due to the computational cost involved in large unit cell calculations. One such study by Huang *et al.* on ZGNRs with *systematic* edge-vacancy defects predicted a loss of magnetism at 33% edge-vacancy concentration [5]. Using a generalized tight-binding (GTB) model with mean-field Hubbard-U that reproduces DFT trends, we were able to verify that the loss of magnetism in the systematic study was due to finite size effects [6]. When *random* edge-vacancy disorder was considered within ensemble-averaging in ultra-thin ZGNRs, our results showed a persistent magnetic state and onset of spin-dependent transport for increasing ribbon length, indicating the dominance of the magnetic interaction against disorder.

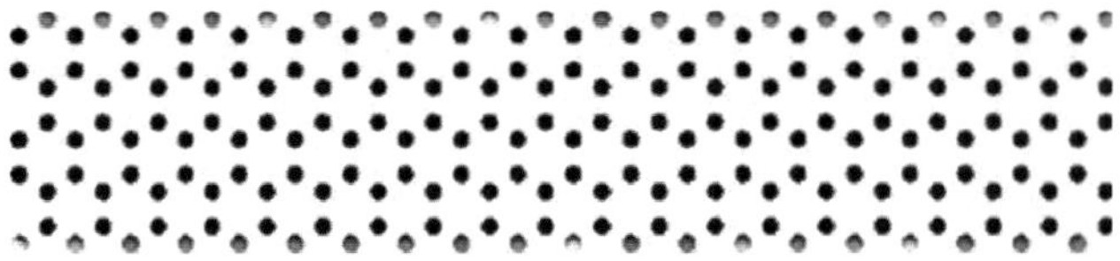

Figure 1. A section of an ultra-thin, 5-atom-wide (9.24 Å) ZGNR device studied in this work. The coloured edge-atoms have weak-disorder implemented by the random perturbation of the site-dependent on-site energy $\epsilon_{i(edge)}$ within the range V=±0.5 eV as applied through the Hamiltonian in Eq. 1. Various system lengths are investigated: 73.8, 86.1, 123.0 and 147.6 Å.

Most tight-binding transport studies on disordered graphene nanoribbons use a non-interacting nearest neighbour (*i.e.,* simple) tight-binding model (STB). For example, Refs. [7, 8]. Here, we extend this work and our previous studies on random edge-vacancy disorder [6], to investigate weak-edge disorder introduced by random perturbation to the local edge-atom chemical potentials in ultra-thin ZGNRs (Fig. 1). Other published studies that use the STB model show ZGNRs to be less robust against this type of disorder compared to armchair graphene nanoribbons (AGNRS), with these systems also having a transport gap that opens due to Anderson localization [7, 8]. By investigating the effect of random weak-edge disorder beyond the STB approximation, we will test the interplay of increased kinetics using an extended tight-binding (ETB) model (up to third nearest neighbour), and interaction effects by applying a generalized tight binding (GTB) model with mean-field Hubbard-U.

2. Theory and computational method

The GTB Hamiltonian is defined as

$$H = \sum_{i\sigma} \epsilon_i n_{i\sigma} n_{i\sigma} - \sum_{ij\sigma} (t_{ij} c^{\dagger}_{i\sigma} c_{j\sigma} + H.c.) + \sum_{i} U n_{i\sigma} n_{i-\sigma} \tag{1}$$

where $c^{\dagger}_{i\sigma}(c_{j\sigma})$ is the fermion creation (destruction) operator, which creates (destroys) an electron with spin $\sigma = \{\uparrow, \downarrow\}$ at site i (j), $n_{i\sigma} = 1$ or 0 is the spin-dependent number operator, and $H.c.$ is the Hermitian conjugate. The model is parameterised using the parameter set in Ref. [9], where $U = 2.0$ eV denotes the Hubbard-U (*i.e.,* the local Coulomb interaction energy between opposite spins [10]), and t_{ij} are the hopping energies equal to 2.7 eV, 0.20 eV and 0.18 eV for first, second and third nearest neighbor hopping, respectively. To model random weak-edge disorder, the local on-site energy of each edge atom $\epsilon_{i(edge)}$ has been randomly perturbed in the range $V = \pm 0.5$ eV (Fig. 1). To solve the Hamiltonian, a mean-field approximation is used maintaining a half-filled system, where

$$n_{i\sigma} n_{i-\sigma} = \langle n_{i\sigma} \rangle n_{i-\sigma} + n_{i\sigma} \langle n_{i-\sigma} \rangle - \langle n_{i\sigma} \rangle \langle n_{i-\sigma} \rangle \tag{2}$$

and $\langle n_{i\sigma} \rangle$ is the expectation associated with the local spin-occupancy. Using this approximation, Eq. (1) can be decoupled and linerarized into two spin-dependent Hamiltonians, which are then solved self-consistently. For further details about the self-consistent method see Ref. [11].

The coherent transport properties of the ZGNRs are calculated using the Landauer-Bütticker formalism [12] assuming that the device has semi-infinite ZGNR leads. The spin-dependent conductance $\mathrm{G}_\sigma(E)$ at energy E is determined from the transmission function $T_\sigma(E)$, such that,

$$\mathrm{G}_\sigma(E) = \frac{e^2}{h} T_\sigma(E) \tag{3}$$

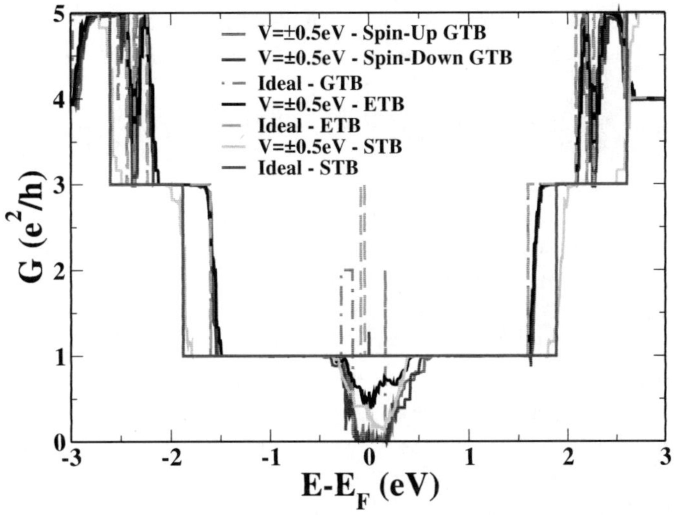

Figure 2. The ensemble-averaged conductance G in units of the quantum conductance (e^2/h) versus the energy E relative to the Fermi energy E_F for ideal and random weak-edge-disordered STB, ETB and GTB systems measured in the range -3 eV $\leq E - E_F \leq 3$ eV. The local on-site potential is randomly varied in the range $V = \pm 0.5$ eV. The systems have a device length of 146.7 Å. For the GTB results, both spin-channels are shown.

where

$$T_\sigma(E) = \mathrm{Tr}[\Gamma_{L\sigma}(E)\mathrm{G}_\sigma^{Ret}(E)\Gamma_{R\sigma}(E)\mathrm{G}_\sigma^{Adv}(E)] \tag{4}$$

and $\mathrm{G}_\sigma^{Ret/Adv}(E)$ are the retarded/advanced Green's functions. The $\Gamma_{L/R\sigma}(E)$ matrices are calculated from

$$\Gamma_{L/R\sigma}(E) = [\Sigma_{L/R\sigma}^{Ret}(E) - \Sigma_{L/R\sigma}^{Adv}(E)] \tag{5}$$

where

$$\Sigma_{L/R\sigma}^{Ret/Adv}(E) = V_{L/R}^{\dagger} g_{L/R\sigma}^{Ret/Adv}(E) V_{L/R}. \tag{6}$$

Here, $V_{L/R}$ describes the coupling between the ZGNR device and the L/R lead, and $g_{L/R\sigma}^{Ret/Adv}(E)$ are the retarded/advanced surface Green's functions for the leads, which have been obtained using the decimation iteration method [13]. The transmission function is calculated using the methods described in Ref. [14]. Our application of this formalism was tested against density functional theory transport results in Hancock *et al.* 2010 [9], and also applied in other works, such as Baldwin and Hancock 2016 [6]. To calculate the effect of random edge-disorder, an ensemble average is obtained where $\chi \approx \bar{x} = \frac{1}{N}\sum_{i=1}^{N} x_i$. Here, $\bar{x}$ is the average value of a calculated property taken over the sample size, N, x_i is the measured property for the *ith* system in the ensemble population, and χ is the ensemble average. The standard error (SE) associated with the ensemble average is SE $= \frac{\sigma}{\sqrt{N}}$ where σ is the standard deviation.

3. Results & Discussion

Figure 2 shows the ensemble-averaged conductance for the disordered and ideal STB, ETB and GTB systems as a function of the energy E relative to the Fermi energy E_F. A device length of 146.7 Å was chosen as longer devices are more susceptible to disorder due to the expected decrease in conductance arising from increased charge-carrier localization. To determine the number of randomly-defected systems (N) in each ensemble that are needed to obtain these

IOP Conf. Series: Journal of Physics: Conf. Series **964** (2018) 012001 doi:10.1088/1742-6596/964/1/012001

results, a convergence test was performed for the ensemble-averaged conductance measured at the Fermi energy G_{E_F} in the STB, ETB and GTB calculations. In performing this test, a strict convergence criterion was set such that the ensemble-averaged results were deemed converged when $\mathrm{G}_{E_F}[N] - \mathrm{G}_{E_F}[N{-}1] \leq 0.01\ e^2/h$. Application of this criterion resulted in convergence of the transport results at N=24 for the non-interacting STB and ETB systems, and at N=12 for the GTB spin-up and spin-down solutions. The smaller number of N=12 systems required to converge the ensemble-average GTB transport results, compared to N=24 for the ETB and STB solutions, indicates the stabilizing effect of the mean-field Hubbard-U against disorder.

The ensemble-averaged transport results show a reduced conductance for the disordered systems, particularly about E_F, compared to the results for the ideal devices (Fig. 2). In the non-interacting results, the ETB system is found to be more robust to disorder compared to the STB system due to the increased kinetic degrees of freedom in the former, with both systems exhibiting perturbations to the conduction up to the second conduction step. Although the conduction is reduced at E_F for the disordered STB and ETB systems, it remains finite (hence statistically significant) against the calculated standard error (SE) (Table 1)—*i.e.*, these results are greater in magnitude than the $0.01{*}e^2/h$ threshold established for experimental validity [1].

The ensemble-averaged GTB results for the interacting disordered system show a break-down in the Hubbard-U gap when compared to the ideal GTB solutions (Fig. 2). A finite conductance is determined for the spin-up channel, and a zero conductance for the spin-down channel at E_F against the SE and the $0.01{*}e^2/h$ criterion (Table 1). The asymmetry of the disorder across the ribbon-width is the underlying cause of these spin-dependent results. The unequal local chemical potentials on the device edge-atoms obtained through the random edge-disorder, as well as at the lead connections, perturb the magnetic state and transport leading to a finite spin-dependent conduction within the Hubbard-U gap region in the GTB solutions.

Table 1. Ensemble-averaged G_{E_F} and associated standard error (SE) for the disordered systems.

System	G_{E_F} (e^2/h)	SE (e^2/h)	$\mathrm{G}_{E_F} \pm$ SE (e^2/h)
STB	0.23	0.09	0.14 to 0.32
ETB	0.5	0.1	0.4 to 0.6
GTB (spin-up channel)	0.05	0.02	0.03 to 0.07
GTB (spin-down channel)	8E-5	3E-5	5E-5 to 1.1E-4

To determine the effect of charge carrier localization due to disorder, the conductance at E_F is plotted as a function of the system length (Fig. 3). At a given energy E, the localization length ζ can be obtained from

$$\mathrm{G}(L) = \mathrm{G}_0(L) e^{\frac{-L}{\zeta}}, \tag{7}$$

where G is the conductance in the disordered system, G_0 is the conductance of the ideal system and L is the length of the device [15]. Using this equation, and from the exponential fits to the results in Fig. 3, the localization lengths have been determined as 65 Å for the STB system and 101 Å for the ETB system, both with ~30% uncertainty, meaning that the charge-carriers are well-localized in the previously studied device length of 146.7 Å. The STB results are found to be in good agreement compared to ensemble-averaged STB results on similarly disordered systems [7], and as expected, the ETB localization length is longer than the STB localization length due to the increased kinetic degrees of freedom in the former. At longer device lengths and/or for higher maximum values of the disorder strength $|V|$, a disorder gap is expected that

IOP Conf. Series: Journal of Physics: Conf. Series **964** (2018) 012001 doi:10.1088/1742-6596/964/1/012001

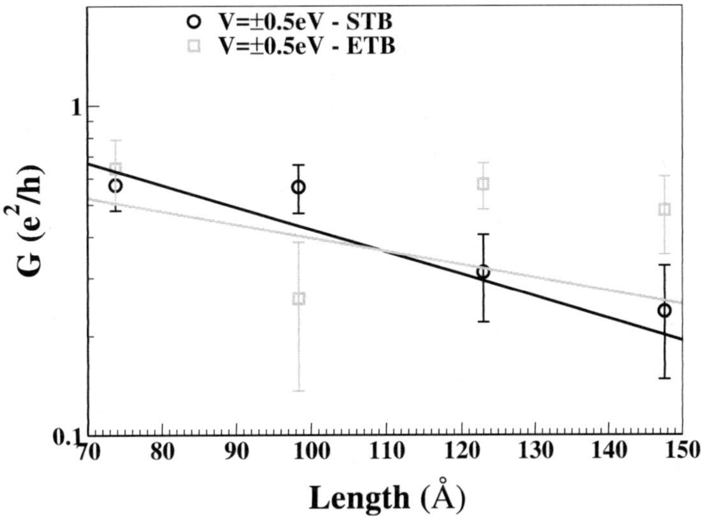

Figure 3. The ensemble-averaged conductance G measured at the Fermi energy E_F in units of quantum conductance (e^2/h) versus system length for the STB and ETB random edge-disordered systems. The local on-site potential is randomly varied in the range $V = \pm 0.5$ eV. Using exponential fitting, the localization lengths are determined as 65 Å for the STB system and 101 Å for the ETB system, both with ~30% uncertainty. The error bars correspond to ± the standard error (SE).

will influence these findings, as well as begin to compete with the Hubbard gap in the GTB results, with the study of these effects being the topic of future work.

4. Conclusion

An introductory study has been performed to gauge the significance of random weak-edge disorder on the coherent transport properties of ultra-thin ZGNRs beyond the simple (*i.e.*, first nearest-neighbour) tight-binding (STB) model. These extensions have included up to third nearest-neighbour hopping in an extended tight-binding (ETB) model, as well as a mean-field Hubbard-U interaction (GTB model). Within ensemble averaging, the effect of the disorder causes charge-carrier localization reducing the STB, ETB and GTB conductances, with the ETB system being more robust relative to the STB due to increased kinetic degrees of freedom. Faster convergence of the ensemble-averaged transport relative to the ensemble size for the GTB also indicates the stabilizing effect of the mean-field Hubbard-U. Asymmetric edge-disorder and connection to the leads results in spin-dependent perturbations that reduce the effective Hubbard-U causing single spin-channel conductance at the Fermi energy. Further studies will involve testing a range of on-site energy disorder strengths and device dimensions.

Acknowledgments

This work made use of the facilities of the N8 HPC Centre of Excellence, provided and funded by the N8 Consortium and EPSRC (United Kingdom), EP/K000225/1. We also acknowledge the University of York's award of a dedicated cluster (Nano) to YH's research group for theoretical graphene research. An EPSRC (United Kingdom), EP/P505798/1, PhD studentship was awarded to YH from the Department of Physics for JPCB. All data created during this research are available by request from the University of York Data Catalogue http://dx.doi.org/10.15124/ea325b7b-c4ed-4bbf-9c4a-0de6e8c0cbba.

References

[1] Han M Y, Özyilmaz B, Zhang Y and Kim P 2007 *Physical Review Letters* **98**, 206805
[2] Allen M J, Tung V C and Kaner R B 2010 *Chemical Reviews* **110**, 132
[3] Son Y -W, Cohen M L, and Louie S G 2006 *Physical Review Letters* **97**, 216803
[4] Tao C, Jiao L, Yazyev O V, Chen Y -C, Feng J, Zhang X, Capaz RB, Tour J M, Zettl A, Louie S G, Dai H and Crommie M F 2011 *Nature Physics* **7**, 616
[5] Huang B, Liu F, Wu J, Gu B -L and Duan W 2008 *Physical Review B* **77** 153411
[6] Baldwin J P C and Hancock Y 2016 *Physical Review B* **94**, 165126
[7] Li T C and Lu S -P 2008 *Physical Review B* **77**, 085408
[8] Cresti A, Nemec N, Biel B, Niebler G, Triozon F, Cuniberti G, and Roche S 2008 *Nano Research* **1**, 361
[9] Hancock Y, Uppstu A, Saloriutta, K, Harju A and Puska M J 2010 *Physical Review B* **81**, 245402
[10] Hubbard J 1963 *Proceedings of the Royal Society (London) Series A* **276**, 238
[11] Hancock Y and Smith A E 2003 *Physica E* **18** 383
[12] Datta S 1997 *Electronic Transport in Mesoscopic Systems* Cambridge University Press, Cambridge, U.K.
[13] Guinea F, Tejedor C, Flores F, and Louis E 1983 *Physical Review B* **28**, 4397
[14] Muñoz-Rojas F, Jacob D, Fernández-Rossier J and Palacios J 2006 *Physical Review B* **74**, 195417
[15] Beenakker C W J 1997 *Reviews of Modern Physics* **69** 731

IOP Conf. Series: Journal of Physics: Conf. Series **964** (2018) 012002 doi:10.1088/1742-6596/964/1/012002

Direct observation of spin polarization in GaAs quantum wires by transverse electron focusing

C. Yan[1,2], S. Kumar[1,2], M. Pepper[1,2], K. Thomas[1,2], P. See[3], I. Farrer[4], D. Ritchie[5], J. Griffiths[5], G. Jones[5]

[1] London Centre for Nanotechnology, 17-19 Gordon Street, London WC1H 0AH, United Kingdom
[2]Department of Electronic and Electrical Engineering, University College London, Torrington Place, London WC1E 7JE, United Kingdom
[3]National Physical Laboratory, Hampton Road, Teddington, Middlesex TW11 0LW, United Kingdom
[4]Department of Electronic and Electrical Engineering, University of Sheffield, Mappin Street, Sheffield S1 3JD, United Kingdom
[5] Cavendish Laboratory, J.J. Thomson Avenue, Cambridge CB3 OHE, United Kingdom

E-mail: uceeya3@ucl.ac.uk

Abstract. We present transverse electron focusing measurements in the two dimensional electrons gas formed at the interface of a GaAs/AlGaAs heterostructure. The experimental arrangement consists of two orthogonal quantum point contacts (QPCs), one acting as injector and the other as detector of the collimated 1D electrons as a function of transverse magnetic field. The focusing spectrum shows anomalous behaviour, the first and third focusing peaks split into two sub-peaks while second peak remains as a single peak. The observed splitting, a signature of spin states, arises from the spin-orbit interaction when the 1D electrons are injected into the 2D regime, thus allowing us to manipulate the spin states within the 1D channel.

There is considerable interest in the spin properties of clean one-dimensional (1D) quantum wires, the science of which has potential for spintronics and spin-based logic devices. Generally electrons in a quantum wire are spin degenerate, but spin polarisation becomes observable on the application of a large in-plane magnetic field[1]. Although spontaneous spin polarisation is forbidden in a strictly 1D system of infinite length, according to the Lieb-Mattis theorem[2], phenomena attributed to spin polarisation[3, 4] such as the 0.7 anomaly[1, 5, 6] and source-drain bias induced 0.25 structure[7] have been observed in quasi-1D systems. A direct measurement of the degree of spin freedom is thus necessary to lead to a comprehensive understanding of these features to complement conductance measurements[8].

A typical transverse electron focusing setup consists of an injector and a detector, generally along a plane such that on application of a small perpendicular magnetic field ($B_{\perp}$) the electrons will exhibit cyclotron motion[9, 10]. Once the cyclotron radius (r_c) matches the condition $N\times 2r_c = L$ (N is an integer and L is the separation between the injector and detector), the injected electrons are guided into the detector and result in a voltage drop across the detector (refered as V_{cc} hereafter). The cyclotron radius is directly proportional to $B_{\perp}$, thus giving rise to focusing peaks periodic in $B_{\perp}$.

It has been predicted theoretically[8] that an imbalance of the spin-split branches, due to the spin-orbit interaction, can be detected using transverse electron focusing by means of observation

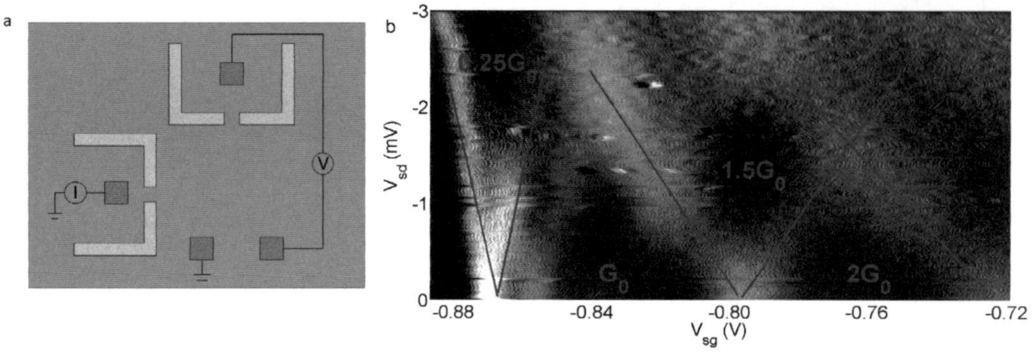

Figure 1. Experiment setup and characteristic of the QPC. a, schematic of the focusing device; the yellow bars are metallic gates and the red blocks are Ohmic contacts. **b**, colour plot of the transconductance $\frac{dG}{dV_{sg}}$ as a function of dc source-drain bias, where V_{sg} is the voltage on the split gates to define the QPC. The bright (dark) regime is for high (low) transconductance. Conductance plateaus quantized in the units of G_0 occur with zero bias voltage while additional structures appear with subsequent increment of the source-drain bias voltage.

of a split in the odd-numbered focusing peaks, where the height of each sub-peak is proportional to the population of detected spin states. It has been confirmed experimentally in a GaAs hole gas[11, 12] and an InSb electron gas[13], the materials with large spin-orbit interaction, that the first focusing peak splits into two sub-peaks where each sub-peak was associated with a spin state.

Here we present results of transverse electron focusing experiments in GaAs based electron gas where a pronounced splitting of the first focusing peak was observed.

The devices were fabricated from a high mobility two dimensional electron gas formed at the interface of GaAs/$Al_{0.33}Ga_{0.67}As$ heterostructure. At 1.5 K, the measured electron density was $1.80\times10^{11}cm^{-2}$ and the mobility was $2.17\times10^{6}cm^{2}V^{-1}s^{-1}$, therefore the mean free path is over 10 μm which is much larger than electron propagation length which is around 2 μm. The experiments were performed in a cryofree dilution refrigerator with an electron temperature of 70 mK, using the standard lockin technique. For the two-terminal conductance (G) measurement, an excitation voltage of 10 μV at 77 Hz was applied whereas for the four-terminal focusing measurement a current excitation of 10 nA at 77 Hz was used[14, 15].

In contrast to the conventional linear focusing device geometry[9, 10] where the central gate is shared between the injector and detector, we used an orthogonal focusing device geometry to allow independent control of the injector and detector as shown in Fig. 1(a). The orthogonal configuration avoids the possible cross-talking between the injector and detector[14]. Both the injector and detector exhibited well defined 1D characteristics as shown in Fig. 1(b). The conductance plateaus were quantized at integer multiple of G_0 ($2e^2/h$) for zero dc source-drain bias, and at half integer plateaus, i.e., 1.5 G_0, 2.5 $G_0\cdots$ etc. at $V_{sd} \approx$ -1.5 mV. A structure at 0.25 G_0 appears at large source-drain bias voltage (< -2 mV) which is similar to the previous report[7].

The focusing result is shown in Fig. 2 with the injector and detector fixed at G_0, respectively. With negative transverse magnetic field the focusing signal is almost zero (result is not shown) while with positive magnetic field the periodic focusing peaks are well defined and the peaks position is in good agreement with calculation according to

$$B_{focus} = \frac{\sqrt{2}\hbar k_F}{eL} \tag{1}$$

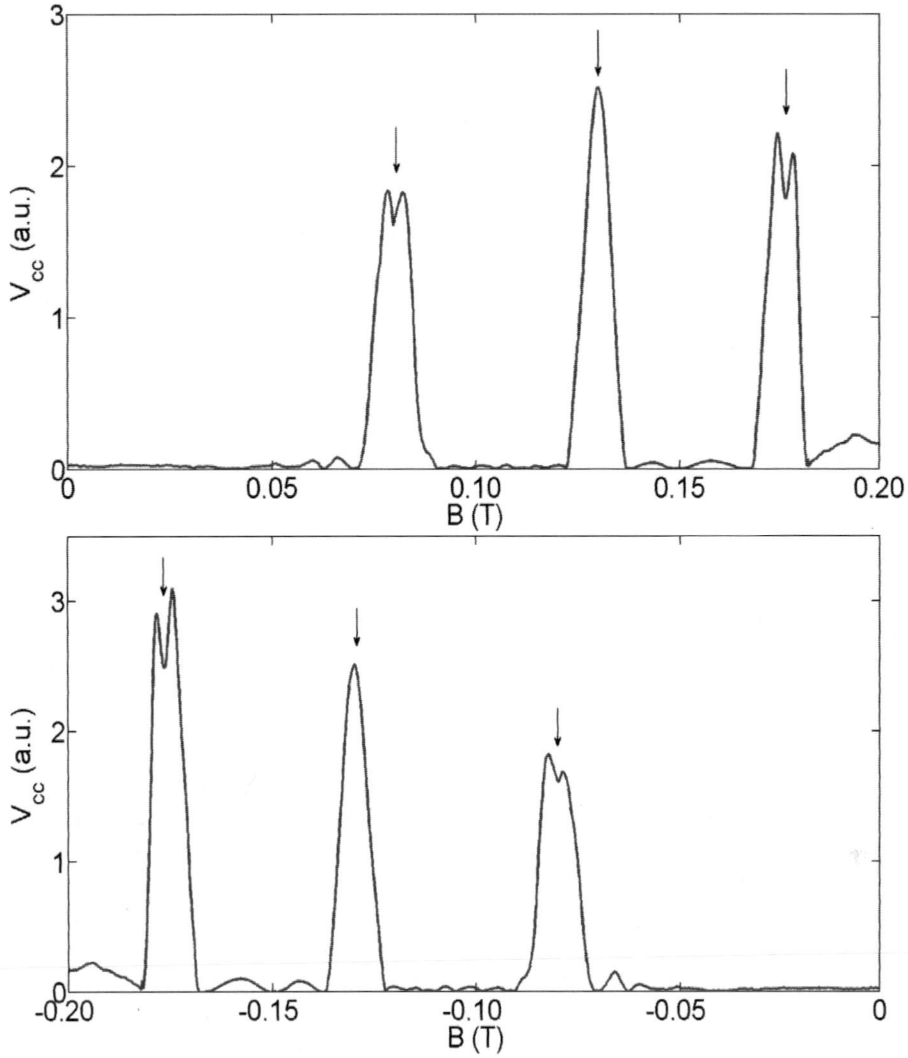

Figure 2. Representative focusing spectrum. Both injector and detector are fixed at first conductance plateau. Periodic focusing peaks are well defined and their position is in good agreement with calculation as highlighted by the arrows. The upper plot shows the result with current flowing from the injector to detector and vice-versa for the lower plot.

A comparison between negative and positive magnetic field results suggests that the Quantum Hall effect and Shubnikov-de Haas (SdH) oscillations are negligible in the regime of focusing[9, 14, 15], and all the features are due to transverse electron focusing only. The first and third focusing peaks split into two sub-peaks, on the other hand, the second focusing peak remains as a single peak. These observations are similar to that reported for p-type GaAs and n-type InSb[11, 12, 13], however, the splitting of first focusing peak (around 6 mT) is much smaller compared to p-type GaAs (around 36 mT) and n-type InSb (around 60 mT), which is consistent with the fact that the energy difference between different spin branches in n-type GaAs is much smaller compared with materials with strong spin-orbit interaction[11, 12, 13]. The focusing spectrum remains qualitatively the same after swapping the role of injector and detector as

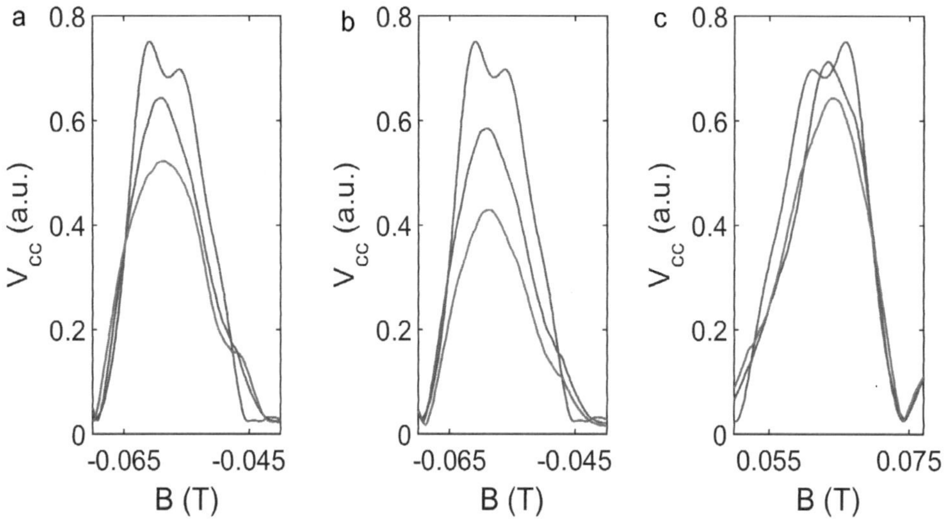

Figure 3. First peak at higher conductance. a, injector conductance was set to G_0 (blue), $3G_0$ (green) and $4G_0$ (red), respectively. Detector was fixed at G_0. **b**, detector conductance was set to G_0 (blue), $3G_0$ (green) and $4G_0$ (red), respectively while the injector was fixed at G_0. **c**, result after swapping the polarity of the magnetic field. Injector conductance was fixed at G_0 (blue), $3G_0$ (green) and $4G_0$ (red). Detector was fixed at G_0.

shown in lower panel of Fig. 2. The result shows that the effect is reproducible and free from impurity after the swap of injector and detector. We also illuminated the device with a red LED and the effect was reproducible with a noticeable change in focusing peaks position due to an increase in carrier concentration. We realise that the odd-even peak splitting arises from the spin-orbit interaction[8, 14, 15, 16]. In addition, the observation was further confirmed to be spin related via in-plane magnetic field dependence study where the splitting of first focusing peak was enhanced with increasing in-plane field[15].

We noticed that the first focusing peak showed a pronounced splitting when the injector conductance was smaller than $2G_0$ ($G_0 = \frac{2e^2}{h}$), however, such splitting was absent at large injector conductance value (e.g. $3G_0$ and $4G_0$) and only a single peak was observed as shown in Fig. 3(a). Similar result was obtained by fixing the injector conductance at G_0 and tuning the detector conductance (Fig. 3(b)) and swapping the polarity of the magnetic field (and the role of injector and detector, Fig. 3(c)). It is also important to emphasize that a single peak at large injector conductance value aligns with the dip between the two sub-peaks rather than one of the sub-peaks. This observation suggests the result is disorder free and does not arise from the shape of wavefunction within the injector. Assuming the peak splitting is due to disorder, then with an even larger angular spreading at larger injector conductance the splitting should persist[17].

In conclusion, we have observed a split in the odd-numbered focusing peaks in GaAs electron gas using transverse electron focusing measurement. The result provides a direct method of probing the spin polarisation in 1D electrons thus opens new opportunities for spintronics and other quantum schemes.

The work was supported by Engineering and Physical Sciences Research Council (EPSRC), UK.

References

[1] Thomas K. J., Nicholls J. T., Simmons M. Y., Pepper M., Mace D. R. and Ritchie D. A.1996 *Phys. Rev. Lett.* **77** 135–138
[2] Lieb E. and Mattis D. 1962 *Phys. Rev.* **125** 164
[3] Wang, Chuan-Kui and Berggren, K.-F. 1996 *Phys. Rev. B* **54** R14257
[4] Wang, Chuan-Kui and Berggren, K.-F. 1998 *Phys. Rev. B* **57** 4552
[5] Yan C., Kumar S., Pepper M., See P., Farrer I., Ritchie D., Griffiths J. P. and Jones G. A. C. 2017 *Phys. Rev. B* **95** 041407
[6] Yan C., Kumar S., Pepper M., See P., Farrer I., Ritchie D., Griffiths J. P. and Jones G. A. C. 2017 *Phys. Rev. Applied* **8** 024009
[7] Chen T. M., Pepper M., Farrer I., Jones G. A. C. and Ritchie D. A. 2012 *Phys. Rev. Lett.* **109** 177702
[8] Reynoso A., Usaj G. and Balseiro C. A. 2007 *Phys. Rev. B* **75** 085321
[9] van Houten H., Beenakker C. W. J., Williamson J. G., Broekaart M. E. I., van Loosdrecht P. H. M., van Wees B. J., Mooij J. E., Foxon C. T. and Harris J. J. 1989 *Phys. Rev. B* **39** 8556-8575
[10] Potok R. M., Folk J. A., Marcus C. M. and Umansky V. 2002 *Phys. Rev. Lett.* **89** 266602
[11] Rokhinson L. P., Pfeiffer L. N., and West K. W. 2006 *Phys. Rev. Lett.* **96** 156602
[12] Chesi S., Giuliani G. F., Rokhinson L. P., Pfeiffer L. N. and West K. W. 2011 *Phys. Rev. Lett.* **106** 236601
[13] Dedigama A. R., Deen D., Murphy S. Q., Goel N., Keay J. C., Santos M. B., Suzuki K., Miyashita S. and Hirayama Y. 2006 *Physica E: Low-dimensional Systems and Nanostructures* **34** 647-650
[14] Yan C., Kumar S., Thomas K., Pepper M., See P., Farrer I., Ritchie D., Griffiths J. P. and Jones G. A. C. 2017 *Appl. Phys. Lett.* **111** 042107
[15] Yan C., Kumar S., Pepper M., See P., Farrer I., Ritchie D., Griffiths J. P. and Jones G. A. C. 2017 *Nanoscale Research Letters* **12**:553
[16] Bladwell S. and Sushkov O. P. 2015 *Phys. Rev. B* **92** 235416
[17] Maryenko D., Ospal F., v. Klitzing K., Smet J. H., Metzger J. J., Fleischmann R., Geisel T. and Umansky T. 2012 *Phys. Rev. B* **85** 195329

IOP Conf. Series: Journal of Physics: Conf. Series **964** (2018) 012003 doi:10.1088/1742-6596/964/1/012003

Cavity assisted spin reconfiguration in a quantum wire

C. Yan[1,2], S. Kumar[1,2], M. Pepper[1,2], P. See[3], I. Farrer[4], D. Ritchie[5], J. Griffiths[5], G. Jones[5]

[1] London Centre for Nanotechnology, 17-19 Gordon Street, London WC1H 0AH, United Kingdom
[2]Department of Electronic and Electrical Engineering, University College London, Torrington Place, London WC1E 7JE, United Kingdom
[3]National Physical Laboratory, Hampton Road, Teddington, Middlesex TW11 0LW, United Kingdom
[4]Department of Electronic and Electrical Engineering, University of Sheffield, Mappin Street, Sheffield S1 3JD, United Kingdom
[5] Cavendish Laboratory, J.J. Thomson Avenue, Cambridge CB3 OHE, United Kingdom

E-mail: uceeya3@ucl.ac.uk

Abstract. In the present work, we investigate the source-drain bias spectrum of a hybrid quantum device consisting of a dot-like QPC coupled to an electronic cavity. A singlet-triplet transition manifested as finite-bias anomaly is observed when the cavity is switched on. Besides, we noticed that the 0.7 conductance anomaly is not affected by the cavity, which provides a valuable insight on the origin of the 0.7 conductance anomaly.

1. Introduction

Nanoscale quantum wire, also known as quantum point contact (QPC), defined by applying gate voltage to a pair of split gates[1] has led to some of celebrated results in condensed matter physics. In addition to the quantization of 1D conductance (G)[2, 3], arising from the single-particle picture, which occurs at G $= N \times \frac{2e^2}{h}$ (where N is an integer), QPC is also a versatile platform to investigate many body phenomena such as spontaneous spin polarization[4, 5, 6, 7, 8], the incipient Wigner crystallization[9, 10] and the Kondo effect[11].

The Kondo effect, arising from the coherent spin-flip scattering between the dynamic localized magnetic impurity and conduction electrons, allows an easily accessible approach in manipulating spin configuration[12]. The Kondo effect has led to a series of fascinating results in quantum dots (QD), ranging from the zero-bias conductance anomaly (ZBA) due to single impurity Kondo effect[13] to finite-bias anomaly (FBA) which is a direct manifestation of singlet-triplet transition[14, 15]. Compared its counterpart in QD, the observation and understanding of the Kondo effect especially the multi-impurity system[16, 17] is far from complete in the QPC.

Here we present an evidence on the singlet-triplet Kondo effect in a novel hybrid system consisting of a QPC coupled to an electronic cavity. We show that by carefully engineering the geometry of the QPC and tuning the coupling between the cavity and QPC, anomalous coexistence of ZBA and FBA can be observed.

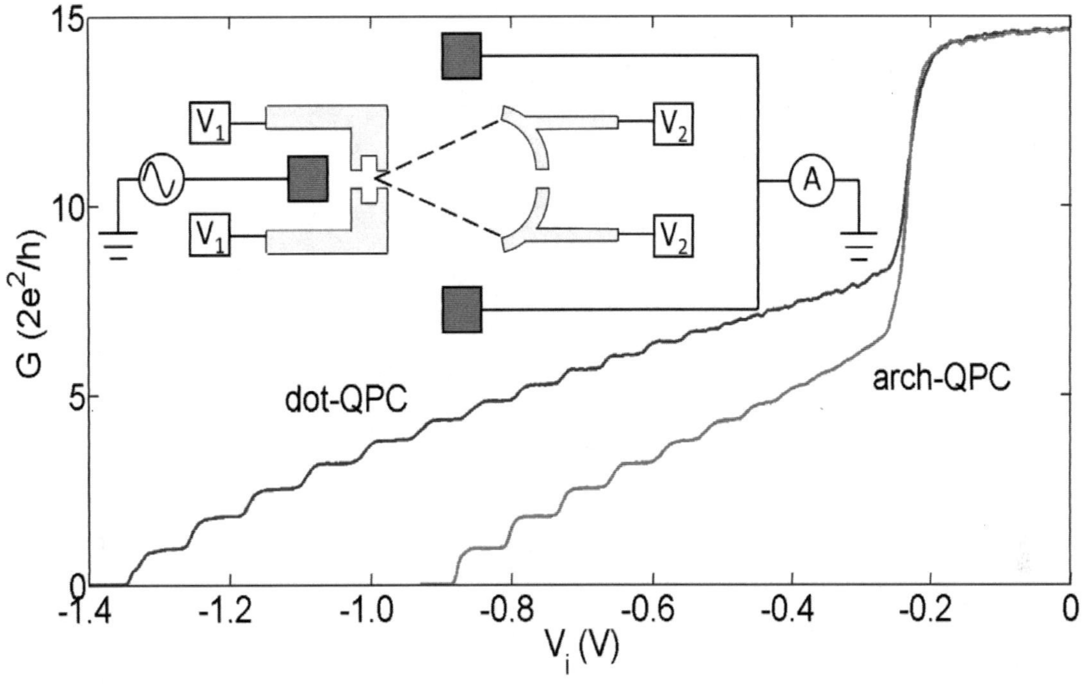

Figure 1. Characteristics of the device and the experiment setup. Main plot shows the conductance characteristics of the dot-QPC and arch-QPC with applied gate voltage V_i (i = 1,2). Inset, the yellow blocks are metallic gates and red squares are Ohmic contacts. The opening angle of the arch is 45° and the radius is 2.0 μm, both the length and width of the QPC embedded in the arch (hereafter referred as arch-QPC) are 200 nm; the length (width) of the injector QPC (named as dot-QPC) is 700 nm (500 nm). The measurement setup in the inset is for the main results, while the conductance characteristics of the individual components are investigated with the typical two-terminal measurement.

2. Experimental detail

The hybrid devices were fabricated on a high mobility two-dimensional electron gas (2DEG) formed at the interface of $GaAs/Al_{0.33}Ga_{0.67}As$ heterostructure. The metallic gates are deposited on the surface which is 90 nm away from the 2DEG. The electron density (mobility) measured at 1.5 K was 1.80×10^{11}cm^{-2} (2.1×10^{6}cm^{2}V^{-1}s^{-1}) therefore the mean free path was over 10 μm which is much larger than the distance between the QPC and reflector (2.0 μm). All the measurements were performed with the standard two-terminal lock-in technique, using an ac voltage of 10 μV at 77 Hz, in a cryofree dilution refrigerator with a lattice temperature of 20 mK (the electron temperature was around 70 mK estimated from the temperature dependence of the Shubnikov-de Haas oscillations).

3. Results and discussion

Both the dot-QPC and arch-QPC show well defined conductance plateaus when measured individually as shown in Fig. 1. It has been carefully examined that the dot-QPC does not show a quantum-dot like behaviour. However, the rectangular grooves of the dot-QPC affect the results significantly in two aspects: first, the reflection from the edge of the split gates could result in the Friedel oscillations which in turn contribute to the formation of an emergent

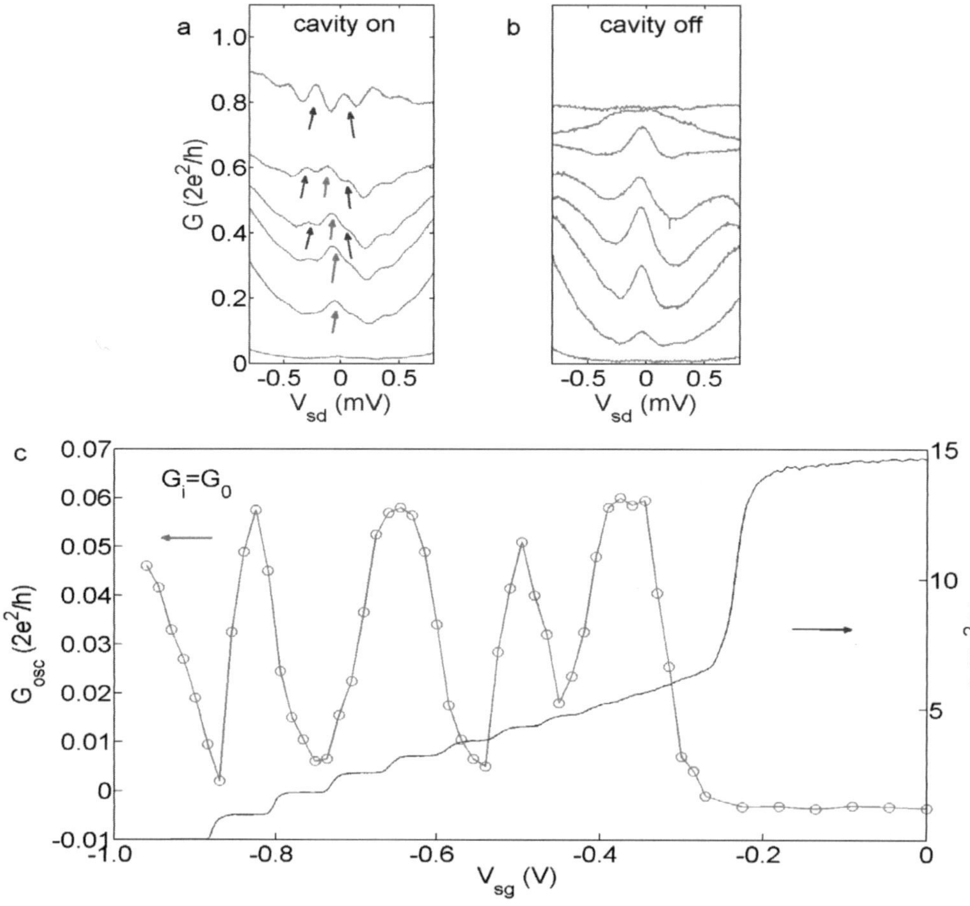

Figure 2. Zero-bias spectrum as a function of the dot-QPC conductance. **a** and **b** show the result with cavity switched on and off, respectively. **c**, conductance oscillation as a function of arch-QPC gate voltage with dot-QPC fixed at G_0.

localized state within the quasi-1D channel; second, coupling between the dot-QPC and the electronic cavity can be tuned in a similar manner to that between a QD and a cavity[19].

The source-drain bias spectroscopy shows dramatic difference when the cavity is switched on and off as shown in Fig. 2. A standard sharp ZBA is observed with the cavity switched off. The ZBA has been studied extensively in QPCs and has been shown to arise from the Kondo effect[11]. Interestingly, more features are observed when the cavity is switched on. In the low conductance regime, the ZBA still dominates the spectrum. With further increasing the conductance of the dot-QPC, FBA peaks occur at ± 0.2 mV in additional to the ZBA (see Fig. 2(b)), which is similar to that observed in a QD[15]. In the large conductance regime (G $\geq 0.8G_0$), only the FBA survives while the ZBA alters into a dip; the FBA-only behaviour, a signature of singlet-triplet transition, is similar to that reported previously[16, 17]. In the presence of in-plane magnetic field, it was found that the ZBA gradually split into two peaks whereas the FBA initially shift towards larger V_{sd} and then smeared out (results not shown here).

Also, it is interesting to notice that, as shown in Fig. 2(c), the oscillatory part of the total conductance G_{osc} arising from the quantum correction due to the cavity (equals to the total

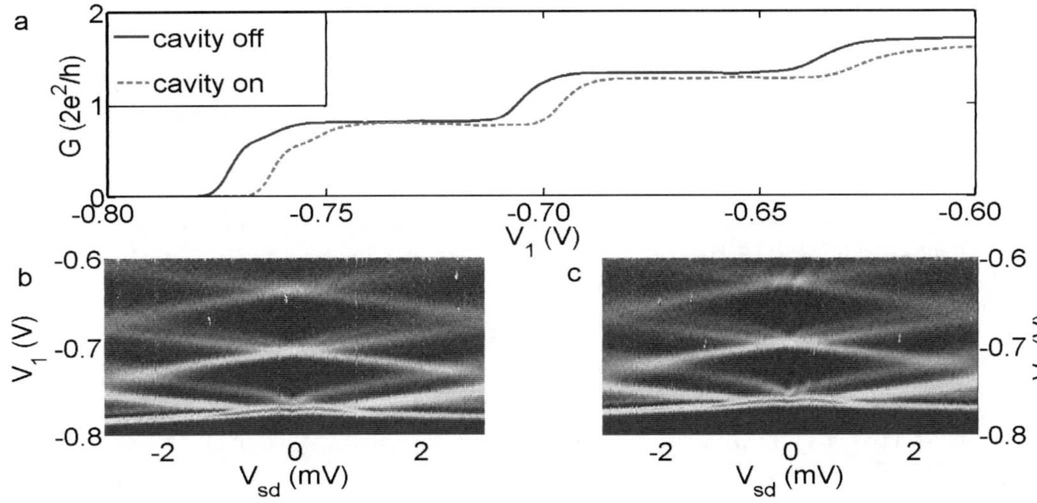

Figure 3. Comparison of the 0.7 anomaly when cavity is switched on and off in sample B. a, injector conductance measured with cavity switched on (red dotted trace) and off (blue solid trace). **b** and **c**, transconductance $\frac{dG}{dV_1}$ with cavity off and on, respectively.

conductance measured with the setup in Fig. 1 minus the dot-QPC conductance; in the plot, dot-QPC was set to G_0) was triggered when the cavity was switched on ($V_{sg} \leq$ -0.25 V, where the electrons underneath the arch-shaped gate were depleted and a quasi-1D channel started to form within the arch-QPC). The oscillatory conductance was still observed even after the arch-QPC was fully pinched-off ($V_{sg} \leq$ -0.85 V) which a signature of coupling between the dot-QPC and cavity[18].

In the previous reports, it was suggested that the occurrence of FBA is correlated with the 0.7 conductance anomaly[16, 17], in other words, the 0.7 anomaly is related to the localized state in the QPC. However, we found no signature of such correlation as shown in Fig. 3(a). It is seen that despite a change in the pinched-off voltage there is no appreciable change in the dot-QPC conductance (including the subband spacing) especially the shape of the 0.7 anomaly. Fig. 3(b) and (c) show the transconductance $\frac{dG}{dV_1}$ with the cavity being switched off and on, respectively, and further indicate that the occurrence of FBA is not related to the 0.7 conductance anomaly. Due to a weak coupling between the dot-QPC and electronic cavity at zero source-drain bias, the states within the 1D channel are not affected significantly. We note that a recent scanning gate microscopy experiment also indicated that the 0.7 anomaly is not correlated with the localized state[21].

In conclusion we have demonstrated a singlet-triplet transition with the assistance of an electronic cavity coupled to a QPC. The singlet-triplet transition manifests itself as a pair of finite bias peaks. Besides, our experiments also reveal that the occurrence of 0.7 conductance anomaly is not correlated with FBA. The hybrid device presented here provides an easily accessible route to spintronics and related quantum schemes.

Acknowledgments
This work was supported by the Engineering and Physical Sciences Research Council (EPSRC), UK.

References

[1] Thornton T. J., Pepper M., Ahmed H., Andrews D., Davies G. J. 1986 *Phys. Rev. Lett.* **56** 1198-1201
[2] Wharam D. A., Thornton T. J., Newbury R., Pepper M., Ahmed H., Frost J. E. F., Hasko D. G., Peacock D. C., Ritchie D. A. and Jones G. A. C. 1988 *Journal of Physics C: Solid State Physics* **21** L209-L214
[3] van Wees B. J., van Houten H., Beenakker C. W. J., Williamson, J. G., Kouwenhoven L. P., van der Marel D. and Foxon C. T. C. 1988 *Phys. Rev. Lett.* **60** 848–850
[4] Thomas K. J., Nicholls J. T., Simmons M. Y., Pepper M., Mace D. R. and Ritchie D. A.1996 *Phys. Rev. Lett.* **77** 135–138
[5] Yan C., Kumar S., Thomas K., Pepper M., See P., Farrer I., Ritchie D., Griffiths J. P. and Jones G. A. C. 2017 *Appl. Phys. Lett.* **111** 042107
[6] Yan C., Kumar S., Pepper M., See P., Farrer I., Ritchie D., Griffiths J. P. and Jones G. A. C. 2017 *Phys. Rev. B* **95** 041407
[7] Yan C., Kumar S., Pepper M., See P., Farrer I., Ritchie D., Griffiths J. P. and Jones G. A. C. 2017 *Phys. Rev. Applied* **8** 024009
[8] Yan C., Kumar S., Pepper M., See P., Farrer I., Ritchie D., Griffiths J. P. and Jones G. A. C. 2017 *Nanoscale Research Letters* **12**:553
[9] Hew W. K., Thomas, K. J., Pepper M., Farrer I., Anderson D., Jones G. A. C. and Ritchie D. A. 2009 *Phys. Rev. Lett.* **102** 056804
[10] Kumar S., Thomas K. J., Smith L. W., Pepper M., Creeth G. L., Farrer I., Ritchie D., Jones G. and Griffiths J. 2014 *Phys. Rev. B* **90** 201304
[11] Cronenwett S. M., Lynch H. J., Goldhaber-Gordon D., Kouwenhoven L. P., Marcus C. M., Hirose, K., Wingreen N. S. and Umansky V. 2002 *Phys. Rev. Lett.* **88** 226805
[12] Kondo J. 1964 *Progress of theoretical physics* **32** 37–49
[13] Cronenwett S. M., Oosterkamp T. H. and Kouwenhoven L. P 1998 *Science* **281** 540–544
[14] Paaske J., Rosch A., Wölfle P., Mason N., Marcus C. M. and Nygård J. 2006 *Nature Physics* **2** 460–464
[15] Roch N., Florens S., Bouchiat V., Wernsdorfer W. and Balestro F. 2008 *Nature* **453** 633–637
[16] Iqbal M. J., Levy R., Koop E. J., Dekker J. B., De Jong J. P., van der Velde J. H. M., Reuter D., Wieck A. D., Aguado, R., Meir Y. and van der Wal C. H. 2013 *Nature* **501** 79–83
[17] Brun B., Martins F., Faniel S., Hackens B., Bachelier G., Cavanna A., Ulysse C., Ouerghi A., Gennser U., Mailly D., Huant S., Bayot V., Sanquer M. and Sellier H. 2014 *Nature Communication* **5** 4290
[18] Saito M., Usuki T., Okada M., Futatsugi T., Kiehl R.A. and Yokoyama N. 1994 *Appl. Phys. Lett.* **65** 3087-3089
[19] Rössler C., Oehri D., Zilberberg O., Blatter G., Karalic M., Pijnenburg J., Hofmann. A., Ihn T., Ensslin K., Reichl C. and Wegscheider W. 2015 *Phys. Rev. Lett.* **115** 166603
[20] Jaksch P., Yakimenko I. and Berggren K. F. 2006 *Phys. Rev. B* **74** 235320
[21] Iagallo A., Paradiso N., Roddaro S., Reichl C., Wegscheider W., Biasiol G., Sorba L., Beltram F. and Heun S. 2015 *Nano Research* **8** 948–956

Investigation of properties of nanobridge Josephson junctions and superconducting tracks fabricated by FIB

B Li[1,2], T Godfrey[1,3], D Cox[1,2], T Li[1,3], J Gallop[1], S Galer[1], A Nisbet[2], Ed Romans[1,3] and L Hao[1]

[11]National Physical Laboratory, Hampton Rd., Teddington TW11 0LW, UK2
[2]University of Surrey, Guildford, Surrey GU2 7XH, UK
[3]London Centre for Nanotechnology, University College London, WC1H 0AH, UK

E-mail: ling.hao@npl.co.uk

Abstract. An important requirement across a range of sensitive detectors is to determine accurately the energy deposited by the impact of a particle in a small volume. The particle may be anything from a visible photon through to an X-ray or massive charged particle. We have been developing nanobridge Josephson junctions based SQUIDs and nanoSQUID devices covering the entire range of particle detection energies from 1eV to MeV. In this paper we discuss some developments in nanobridge Josephson junctions fabrication using focussed ion beam (FIB) and how these developments impact future applications. We focus on tuning of the transition temperature of a superconducting thin-film absorber, with the aim to match the absorber T_c to the working temperature range of the SQUID and also on using a new Xe FIB to improve Josephson junction and superconducting film quality.

1. Inductive transition edge detector

The bolometer systems we are developing are based on an inductive superconducting transition edge detector (ISTED). The basic principle [1] is that a superconducting thin film absorber is deposited within the loop of a micrometre-scale planar SQUID [2]. Figure 1 shows an SEM image of a square loop SQUID with two nanobridge Josephson junctions, with a thin film Nb absorber deposited within the SQUID loop. The absorber is selected so that it has a lower transition temperature T_c than the SQUID. If the temperature of the chip is adjusted so that both the SQUID is in its optimum operating range and the absorber is just below its own transition temperature, the absorber is in the regime where the superconducting penetration depth (λ) is a very sensitive function of temperature. If the absorber receives some energy input from an incident energetic particle the energy deposited in it will raise its temperature, thus increasing the superconducting penetration depth and resulting in a change in the effective area of the SQUID loop. If the SQUID is biased with an offset flux the change in flux periodicity will result in a large change of SQUID output voltage, even for a change in temperature as small as 1μK.

The superconducting thin-film absorber layer can be seen as a transducer, converting the thermal energy signal into magnetic flux change, which is detected by the surrounding SQUID as a pulse in its voltage output. Unlike conventional superconducting transition edge sensors (TES) [3], the ISTED absorber is not biased into the normal state, remaining permanently superconducting so that the system is not subject to Nyquist noise, thereby improving its overall energy sensitivity. By adding to the

IOP Conf. Series: Journal of Physics: Conf. Series **964** (2018) 012004 doi:10.1088/1742-6596/964/1/012004

absorber a further thin layer of a material (such as amorphous carbon or graphene) an ISTED can be converted to a microbolometer or a micro/nanodosimeter [4], where the absorbing layer is selected for thickness and other properties to optimize response for the chosen application. Once an incident event happens, the energy released will be absorbed by the absorber, resulting in a pulse signal of the output voltage. The magnitude of the pulse signal quantitatively reflects the energy released in this event.

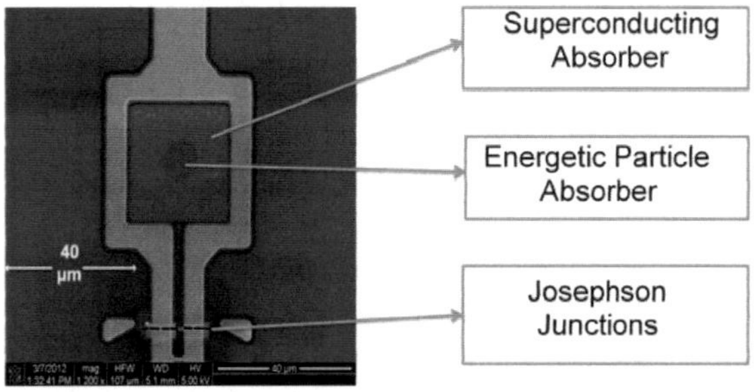

Figure 1. An SEM image of a prototype micron-scale ISTED device with tissue-equivalent absorber aimed for micro/nano dosimetry.

2. Tuning the SQUID-Absorber Transition Temperature

For optimal performance of the ISTED calorimeter the T_c of the absorber film should be slightly lower (typically 1K) than the T_c of the SQUID itself. This means that the Nb nanobridge SQUID which is usually operated at around 2K below the T_c of pure Nb has an absorber which is still superconducting at the SQUID operating temperature but remains in the region where its penetration depth is still strongly temperature dependent. We have investigated two different methods to 'tune' the T_c of the absorber film. The first involves exposure to a scanning Gallium (Ga) FIB of which the amount of Ga ions deposited into a unit area of the Nb film can be easily and precisely controlled. Of course this process not only leads to reduction of the film thickness but also causes damage to the thin film's crystal lattice and implantation of Ga ions. After a selected number of passes, corresponding to specific exposure, the T_c of the film is re-measured then the process is repeated. Another technique for tuning the T_c of a film is to control the thickness during the deposition of the film. We have deposited Nb films with different thicknesses and measured the transition temperature. The first method, by FIB dosing and milling, only requires a single deposition and fabrication step for both SQUID and absorber. The second method, controlling the deposited thickness of the absorber, requires separate Nb deposition and patterning steps, but may provide greater control over T_c. The two methods are described below.

2.1. Gallium ion dosing and milling

When energetic ions impinge on a sample surface there are two pronounced effects. Firstly, the ion implants into the sample causing a cascade of damage as it loses energy before coming to rest. Second, if sufficient energy to overcome the binding energy of surface atoms is released by the ion, sample atoms can be ejected in a process known as sputtering [5]. FIB instruments with Ga (and more recently Xe) ion sources exploit this latter phenomena, and are now well established tools in semiconductor and wider material science applications. Less well known is that FIBs can also be used as fabrication and sample modification tools. We use this technique to fabricate nano-SQUID devices where that Ga FIB modifies simple Nb films to produce extremely low noise SQUIDs in a fast two stage process. In these devices the SQUID nanoscale junctions are of the nanobridge where the FIB is used to mill away the film to produce a narrow junction region with suppressed T_c.

Using a similar mechanism to Nb film ion milling, but additionally relying on the Ga ion dosing, a FIB instrument can also be used to tune the T_c of a Nb film. The Nb films were sputter-deposited on a silicon substrate with a very thin Ti adhesion layer and total thickness in the range of 195-200 nm. A four-terminal lead and bond pad arrangement was used to allow the T_c of the film to be measured. The

IOP Conf. Series: Journal of Physics: Conf. Series **964** (2018) 012004 doi:10.1088/1742-6596/964/1/012004

films were exposed with the ion beam across the full width of a track (10 μm) over a length of 3 μm and for a range of ion doses (see Fig.2). Using a 30 kV, 100 pA beam current, with beam step of 10 nm in both x and y direction, and a 1 μs beam dwell at each point gave a dose of 6.2 ions/nm^2 for a single pass of the beam. This dose was repeated in sets of 25 identical passes giving approximately 155 ion/nm^2 per set, up to a maximum of 700 sets. After a certain amount of dose, resistance of the Nb film is measured as a function of temperature and T_c of the dosed Nb film is plotted versus the number of Ga ion beam passes in Figure 3. T_c decreases from 8 K to 4.3 K as the Ga beam pass increases from 100 to 400. The superconductivity of the Nb film is fully destroyed beyond 500 passes.

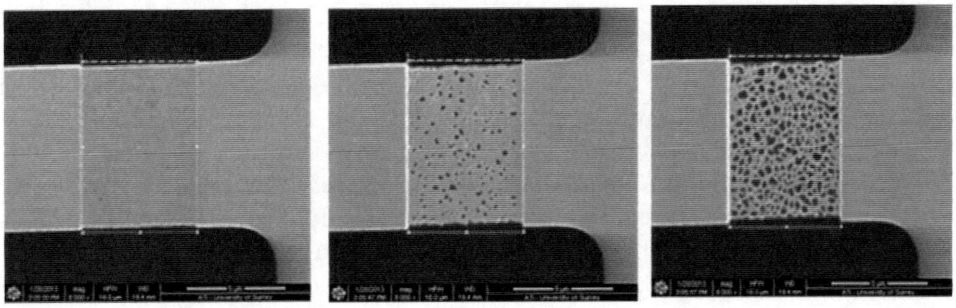

Figure 2. Scanning ion images of Nb film following different levels of ion exposure (corresponding to 100 passes, 400 passes and 600 passes) of Ga ion beams of 30 keV, and beam current ~100 pA. Total removal film is achieved with total ion flux of $4.34 \times 10^9/\mu m^2$. Effects of ion channelling can also be seen. The exposed area is denoted by the yellow dashed box.

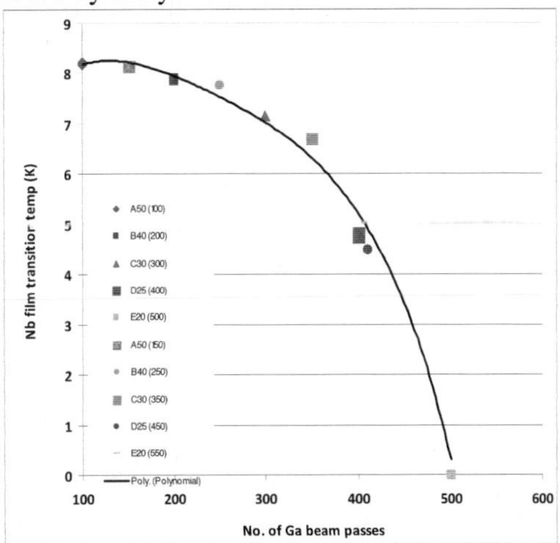

Figure 3. The T_c of the Nb film vs. the No. of Ga ion beam passes. The data is from several different Nb film with an initial thickness of 200 nm. The thickness of film removed in each beam pass is approximately 0.28nm. The solid line shows a third order polynomial fit to the data.

Owing to the importance of ion beam implantation in the semiconductor industry, considerable modelling of ion beam damage and sputtering has been carried out at many levels of sophistication, ranging from molecular dynamics models of single (and few) ion collisions in specific orientations of single-crystals [6], to Monte-Carlo and numerical models approximating to 'bulk' conditions, with ion doses at the upper end of the range offered by FIB milling (>> 10^9 ions/m^2). When considering deposited thin metal films, such as Nb, irradiated with a Ga ion beam, a model of the latter type is more appropriate, as the number of ions can be very large and the samples can be considered as 'bulk' in nature. The main advantages of these 'bulk' models is that at least two of them are offered as readily available, free to use, software packages [7]. Furthermore, in the case of the numerical solution offered by the SUSPRE package it is very rapid and extremely simple to use. The Ga implant profile after one set of 25 repeat passes predicted by the SUSPRE software is shown in Figure 4. As can be seen the predicted profile shows a peak concentration of approximately 15 % Ga, 12nm from the sample surface with a long Gaussian tail extending 40nm into the Nb film. Also the predicted damage profile (not shown) for this dose largely mirrors the implant profile and shows 100% amorphisation just beyond the peak implant concentration. Simulating further repeated passes of the beam at the stated dose shows the damage profile ultimately extending to the full range of the implant (40 nm) with 100% amorphisation after just over 100 repeat passes. Along with the damage profile and implantation range SUSPRE also calculates the expected sputter yield (in Nb atoms per Ga ion) and once again this is in good agreement with SRIM. The predicted yield is almost 3.8 Nb atoms per Ga ion (at 30 kV and normal incidence).

Turning to the actual ion milling, under the stated dose conditions that it takes approximately 700 repeat passes of the beam to fully remove the Nb film to Si substrate. Assuming no change in milling rate this is approximately 0.28 nm of material removed per pass of the beam and with the lattice parameter of Nb being 0.33 nm this approximates to one atomic layer per repeat pass. The model prediction for sputter yield and with a dose of 6.3 ion/nm^2 would give a removal rate of 0.4nm per pass. With the atomic density of Nb approximately 55 atoms per cubic nm and a removal rate of 0.28nm per repeat, the actual sputter yield is therefore 2.5 Nb atoms per Ga ion, only 65% of the calculated rate.

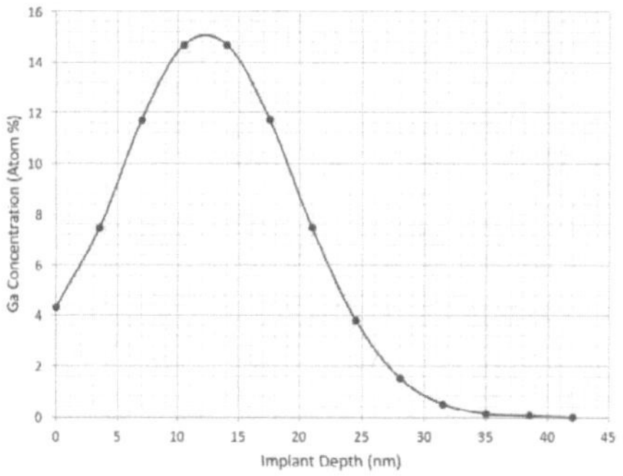

Figure 4. Calculated Ga implant profile for a 30 kV Ga ion dose of approximately 155 ions/nm^2 in Nb.

Assuming a uniform material removal rate, after 200 repeats of the dose the film is thinned by approximately 56 nm. The ion-cascade damaged region would have moved inward into the film uniformly as the film is thinned and with projected range of damage of at least 40nm this would still leave 100 nm of undamaged Nb film. This is the first dose level where we observe a slight suppression in the T_c of the film (Figure 3). Beyond this we see a steady fall in T_c as the film is increasingly dosed to 350 repeats, corresponding to the point the film is approximately half thickness. Allowing for the ion-damaged zone, the undamaged portion of film at this point would be of the order of 60nm thick. Beyond this point further dosing of the film causes a rapid decline in T_c with a further 100 repeat doses causing a larger fall in T_c than the previous 350. At this point the pristine film is approximated to be only 34nm in thickness with any further dosing causing a complete loss of T_c.

2.2. *Use of Xe Focussed Ion Beam*

Recently we have been able to test a different FIB system which uses Xe as the massive ion to provide milling. The potential advantages over a conventional Ga ion beam are that the inert gas will not lead to modified electronic properties of the underlying Nb film since, unlike Ga, the Xe will not form chemical modification of the film through alloying. In addition local electronic doping of the film should be reduced. In other respects the mechanical effects of both Ga and Xe should be similar, in terms of milling rate and damage to the upper un-milled surface layers. The Xe mass is significantly higher than the Ga ion mass which should also somewhat increase the milling rate. We have produced a number of Xe milled Nb nanoSQUIDs based nanobridge junctions and their properties are at least as good as similar Ga devices (see an example I-V curves ad function of the magnetic fields in fig 5).

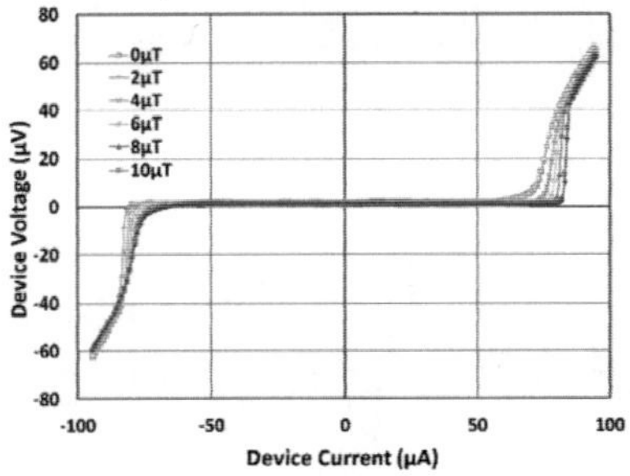

Figure 5. A series of current-voltage curves are shown with a range of magnetic fields.

2.3. *Control of the thickness of the absorber*

A second approach has been investigated as a means of controlling the T_c of a Nb film absorber, to tune it to the optimum operating temperature. Key is that the T_c of a pristine Nb thin film is strongly related to its thickness [8], especially when the thickness is less than 30 nm, allowing the T_c to be tuned by controlling its thickness. In such an approach, the absorber is deposited *after* the fabrication of the SQUID, in a routine lift-off process followed by EBL patterning. The sputtered Nb film thickness is controlled by the sputtering time. The T_c versus thickness relationship is highly dependent on the deposition system and operating conditions. To find out the true thickness of the film with desired T_c, we have deposited several Nb films with different thickness on silicon substrates, and measured the resistance of a patterned track of these films as a function of temperature in a four-probe configuration. In Figure 6 we plot the T_c and corresponding thickness (measured by a Dektak XT surface profiler) of resistance of a patterned track of these films as a function of temperature in a four-probe configuration. Note the T_c increases rapidly with film thickness in the range from 20 nm to 35 nm, which shows in the plot, the trend of T_c as a function of the thickness can be well fitted in a proximity effect model [9, 10]. Since Nb is easily oxidised in air thin Nb films are usually covered by an oxide layer of several nanometres thickness, which can be regarded as normal metal. Due to the existence of such a normal metal layer, the T_c of the Nb film is suppressed by the superconducting proximity effect. The extent of the suppression depends on the thickness ratio of normal layer and superconducting layer.

If we assume the thickness of oxide layer and superconducting layer are *a* and *d* respectively, the T_c of such a two-layer film can be predicted by the relationship

$$T_c = T_{c0}\left(\frac{1.13\Theta_D}{T_{c0}}\right)^{-\frac{N_N(0)}{N_S(0)}\frac{a}{d}}$$

where T_{c0} is the transition temperature of bulk Nb (9.3 K), Θ_D is the Debye temperature of Nb (275 K), and $N_N(0)$ and $N_S(0)$ are the density of states for normal and superconducting layers, respectively. Thus, the thinner the Nb film, the larger the thickness ratio of normal layer and superconducting layer, and therefore the lower the transition temperature T_c.

3. Conclusion and Future Work

The extremely high sensitivity and ultra-low noise performance of nanobridge Josephson junctions make them ideal for fabricating ISTEDs, which can be used to detect rather small energy absorption [11,12]. This makes it practical to try to understand the processes by which the transition temperature of a superconducting thin film may be adjusted for optimal performance, whether it is for SQUIDs/nanoSQUID operation or for tuning the operating temperature of a superconducting absorber. We have been pursuing two distinct methods which have been described in this paper. It is already clear that both are able to provide relatively reproducible changes in T_c, (up to at least 3.5K), whether by means of implantation of ions and damage or by the suppression of T_c by the presence of an underlying normal metal film. Each method will have advantages in different situations.

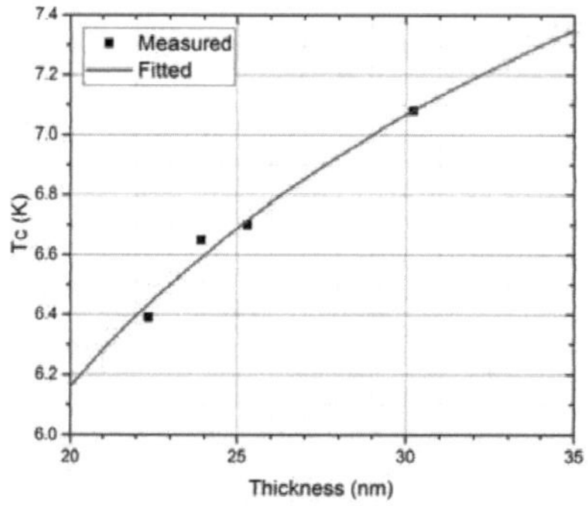

Figure 6. The transition temperature T_c versus the thickness of Nb films. The black dots are measured data and the red curve is fitted by the proximity effect model discussed in the main text.

In parallel with this investigation we have also proceeded to test combined SQUID and absorber designs for prototype ISTED micro/nanodosimeters. The devices will be tested in a cryogen-free pulse tube cooler system with horizontal beam access, to allow a charged particle beam to access the absorber with minimal energy absorption in the mylar windows which support the vacuum necessary for the cryogenic environment. The first beam measurements using the micro-beamline at the National Ion Beam Centre at Surrey University are expected in future. We have also begun investigating the microwave capabilities of these devices, showing that they are likely to be suitable for microwave readout techniques and microwave circuit applications. Microwave reflectometry measurements on inductively coupled SQUIDs are proving to be an advantageous technique for high sensitivity operation. Reflectometry would allow us to make measurements at even higher sensitivity than currently available with microwave inductive readout Johnson noise sources in the junctions will be much reduced.

Acknowledgments
This work was supported by the UK NMS Programme, EPSRC funding and also the NHS i4i funding.

References
[1] Hao L, Macfarlane J, Josephs-Franks P and Gallop J 2003 *IEEE Trans. Appl. Supercond.* **13**, 622
[2] Granata C and Vettoliere A 2016 *Physics Reports* **614** 1-69
[3] Irwin K D and Hilton G C 2005 *Topics Appl. Phys.* **99** 63–150
[4] Furetta C, Prokic M, Salamon R, Prokic V and Kitis G 2001 *Physics Research Section A:* **456** 411-417
[5] Cox D C, Gallop J C and Hao L 2014 *Nanofabrication* **1** 53-64
[6] Yamamura Y and Takeuchi W 1987 *Nuclear Instruments and Methods in Physics Research Section B: Beam Interactions with Materials and Atoms* **29** 461-470
[7] www.srim.org
[8] Bose S, Raychaudhuri P, Banerjee R, Vasa P and Ayyub P 2005 *Phys Rev. Lett.* **95** 147003
[9] Frommberger M, Mattiocco F, Sabon P, Schicke M, Schuster K F and Laborde O 2000 *Proc. 11th Int. Symp. on Space Terahertz Technology*
[10] McMillan W L 1968 *Physical Review* **175** 537
[11] Hao L 2011 *J. Phys.: Conf. Ser.* **286** 012013
[12] J Gallop J, Cox D and Hao L 2015 *Supercond. Sci. Technol.* **28** 084002

Optical Microscopy as a probe of the rate limiting transport lifetime in InSb/$Al_{1-x}In_xSb$ quantum wells

C J McIndo[1,4], D G Hayes[1], A Papageorgiou[1], L A Hanks [1,5], G V Smith[1], C P Allford[1,2], S Zhang[3], E M Clarke[3] and P D Buckle[1]

[1] School of Physics and Astronomy, Cardiff University, Queen's Buildings, The Parade, Cardiff, CF24 3AA, UK
[2] School of Engineering, University of Warwick, Coventry, CV4 7AL, UK
[3] EPSRC National Centre for III-V Technologies, North Campus University of Sheffield, Sheffield, S3 7HQ, UK

E-mail: mcindocj@cardiff.ac.uk

Abstract. Recent reports of magnetotransport measurements of InSb/$Al_{1-x}In_xSb$ quantum well structures at low temperature (3 K) have shown the need for inclusion of a new scattering mechanism not present in traditional transport lifetime models. Observations and analysis of characteristic surface structures using differential interference contrast DIC (Nomarski) optical imaging have extracted representative average grain feature sizes for this surface structure and shown these features to be the limiting low temperature scattering mechanism. We have subsequently modelled the potential profile of these surface structures using Landauer-Büttiker tunnelling calculations and a combination of a Monte-Carlo simulation and Drude model for mobility. This model matches experimentally measured currents and mobilities at low temperatures, giving a range of possible barrier heights and widths, as well modelling the theoretical trend in mobility with temperature.

1. Introduction

Indium antimonide (InSb) exhibits the lowest reported electron effective mass ($m^* = 0.014\, m_e$) [1] and highest reported room-temperature electron mobility ($\mu = 78{,}000\ cm^2V^{-1}s^{-1}$) [1] of any compound semiconductor. These properties make InSb particularly suited to many electronic applications, including low power high frequency electronics and quantum device realisation. There has been recent interest in the development of high quality InSb material following the report of two-dimensional electron gas (2DEG) channel mobilities in excess of 200,000 $cm^2V^{-1}s^{-1}$ at $T = 1.8$ K [2-9] and the recent reports of Majorana fermion observation in InSb nanowires [10, 11]. There is renewed interest in this material system, when coupled with superconducting material, to provide a potentially robust quantum system for information processing.

Furthermore, the strong spin-orbit interaction and extremely large Landé g-factor ($g \approx -50$) [1, 12] exhibited in InSb has gained attention for potential exploitation in spintronics and quantum information control [13-15].

[4] Author to whom all correspondence should be addressed
[5] Present address: Physics Department, Lancaster University, Lancaster, LA1 4YB, UK

IOP Conf. Series: Journal of Physics: Conf. Series **964** (2018) 012005 doi:10.1088/1742-6596/964/1/012005

Previous studies of carrier transport in InSb 2DEGs [3, 4] have considered standard scattering mechanisms using the relaxation time approximation, describing the mobility variation over a wide range of temperature (typically 3 K – 300 K). Whilst there has been good agreement, parameters used have tended to be extreme values to enable acceptable fits to data [9]. It has recently been shown that a previously unaccounted for major scattering mechanism due to surface features (hillocks [6-8]), associated with material quality, must be considered, with this having a major effect on the measured mobility. Considering this additional structural scattering allows for more reasonable values for standard scattering mechanisms, showing that there is immense potential for mobility improvement in this material [8]. Due to this potentially large mobility improvement in InSb QWs, there have been further recent investigations into methods of improved MBE growth via buffers and superlattice structures [6], as well as research into hillock growth and formation [7].

2. Growth and Sample Details

Detailed descriptions of the growth and fabrication of samples can be found in references [8] and [9]. In summary, 30 nm InSb quantum well heterostructures were grown by solid source molecular beam epitaxy (MBE) on semi-insulating GaAs substrates (and so are therefore lattice mismatched with regard to the substrate). Hall bar devices with an aspect ratio of 5:1 (nominally 200 μm × 40 μm) were then fabricated using standard techniques for subsequent magnetotransport measurements.

3. Experimental Determination of Mobility and Sheet Carrier Density

We recently reported observed trends in magnetoresistance Hall measurements of several InSb QW samples with the results reproduced here (see references [8] or [9] for experimental details).

Figure 1 shows the extracted mobilities, μ, and carrier densities, n_{2D}, for a range of different samples at 3 K, with samples with increasing δ-doping (filled symbols) broadly defining three regimes in the data. Initially, in region 1, an increasing mobility is observed for an increase in carrier concentration (from 1×10^{11} cm^{-2} to 3×10^{11} cm^{-2}). This is believed to be due to single subband filling, calculated from Schrödinger-Poisson (S.P.) modelling [16], giving rise to increased Thomas-Fermi screening. The mobility then begins to plateau at ~250,000 cm^{2}V^{-1}s^{-1} for a narrow range of carrier densities (region 2), before decreasing beyond 4×10^{11} cm^{-2} (region 3). In region 3, S.P. modelling shows there is the possibility of multiple subband occupancy and additional intersubband scattering.

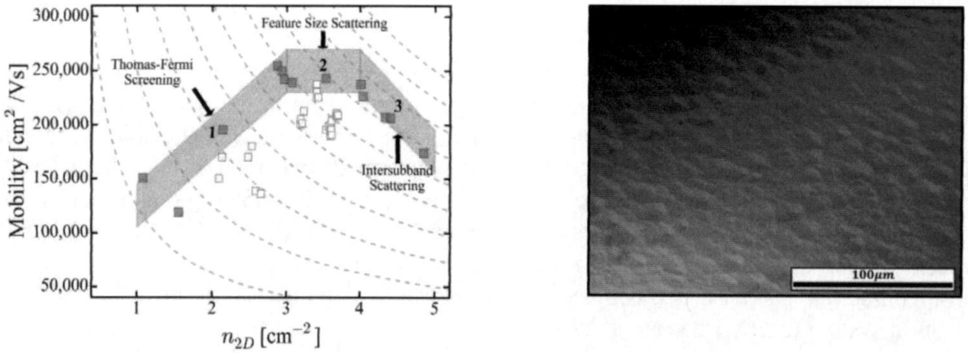

Figure 1. Left: Measured mobility (two carrier fit) vs. 2D sheet carrier density (n_{2D}) from 3 K Hall measurements. A series of samples with increasing δ-doping levels is shown by the larger, filed symbols, broadly defining three regimes (labelled with principal physical cause). Smaller, unfilled symbols indicate similar samples from other growth batches. Dashed lines are contours of constant conductance from 2 mS to 20 mS. Right: Optical Nomarski image of sample surface, magnification × 50, for a sample with n_{2D} ~3×10^{11} cm^{-2} and mobility ~200,000 cm^{2}V^{-1}s^{-1}. The surface is clearly textured, with features of varying sizes.

IOP Conf. Series: Journal of Physics: Conf. Series **964** (2018) 012005 doi:10.1088/1742-6596/964/1/012005

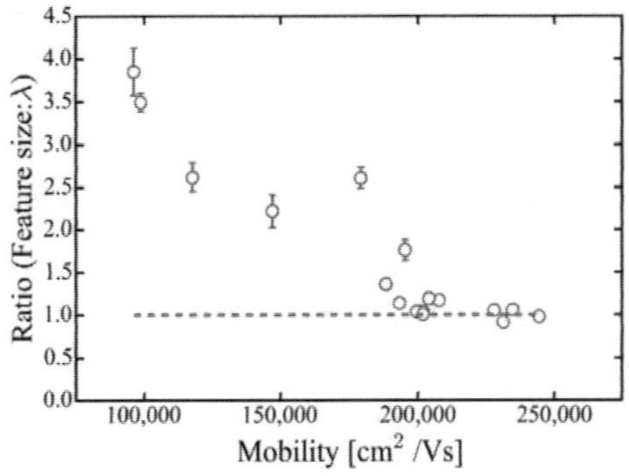

Figure 2. Ratio of mean feature diameter (with one standard deviation error determined from size distributions given by multiple image sampling) to measured mean free patch λ as a function of mobility (open circles), approaching limiting value of 1 (dashed line).

To examine the limiting factors affecting the higher mobility samples, we recently reported on the study of the surface morphology using optical differential interference contrast DIC (Nomarski) imaging [8]. A raw Nomarski image of a standard sample with $n_{2D} \sim 3 \times 10^{11}$ cm^{-2} and $\mu \sim 200{,}000$ cm^{2}V^{-1}s^{-1} is shown in figure 1 at a magnification of $\times$ 50.

The Nomarski image shows clear surface roughness, present similarly on all samples imaged [6, 7]. The roughness consists of approximately circular features with clear boundaries separating features. For the samples studied here, the embedded quantum well where the 2DEG resides is situated at a depth of 50 nm below the surface, whereas the threading dislocations that cause these surface features are created at the substrate/buffer interface at a depth of 3 μm. Any dislocations that propagate through the buffer to the quantum well will therefore continue to the surface. Consequently, due to the proximity of the 2DEG to the surface, it is reasonable to assume that this surface roughness, and in particular the boundaries, have a severe impact on the electron transport in the quantum well.

Using basic image analysis techniques (described in reference [8]), we extracted average feature sizes for multiple samples in figure 1. The ratio of the mean feature diameter to the mean free paths, λ, deduced from 3 K Hall mobility data via a basic Drude transport model, is shown in figure 2. Figure 2 shows there is a clear trend in the ratio of feature size with λ, approaching a value of 1 for the highest mobility samples. This limiting value is strongly suggestive that these features, with an average size of ~2.43 ± 0.13 μm, are the low temperature transport lifetime limiting scattering mechanism. At low temperatures, when phonon effects are reduced, an electron traveling in the quantum well may travel ballistically through a feature until it reaches a boundary where it scatters.

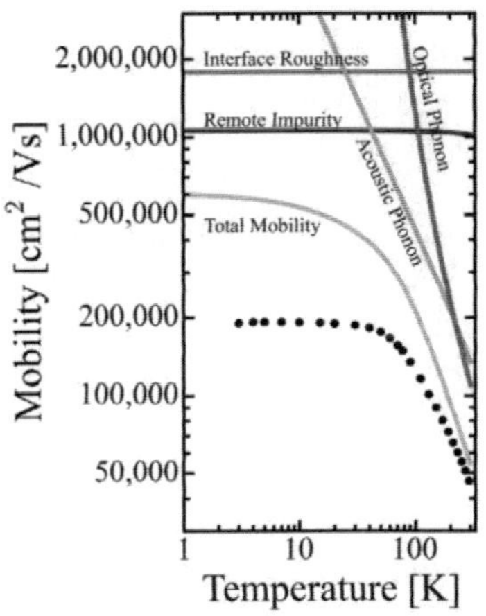

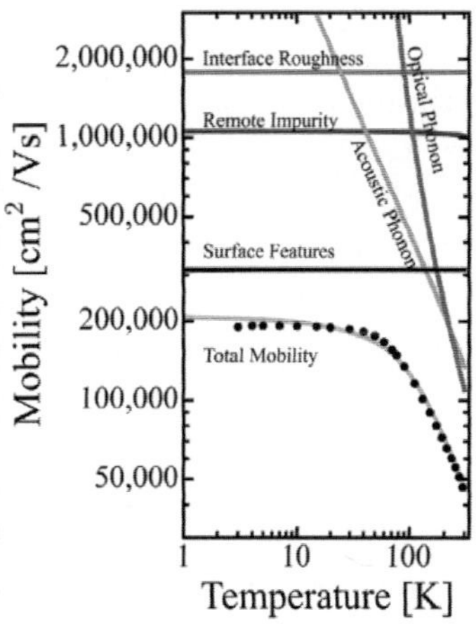

Figure 3. Left: Transport modelled mobility (lines) and measured data (symbols) vs. temperature including standard scattering mechanisms (using typical values [8, 9]). Right: Adapted model including scattering due to 2.43 μm surface features, matching well to Hall measured mobilities.

IOP Conf. Series: Journal of Physics: Conf. Series **964** (2018) 012005 doi:10.1088/1742-6596/964/1/012005

Transport lifetime modelling of InSb QW devices has been performed previously by Orr *et al* [4] and more recently by McIndo *et al* [8] and Hayes *et al* [9]. These models include several of the standard dominant scattering mechanisms present in III-V heterostructures as well as band non-parabolicity, via a modified effective mass, deduced from a six-band Kane model [3, 17, 18].

By consideration for parameters for each sample, including n_{2D} at 3 K, and reasonable values of background p-type impurities (~1 – 10 × 10^{15} cm^{-3}) it was shown [8, 9] that the mobility could not be matched to measured data without the inclusion of surface feature scattering derived from a Drude model, given by

$$\frac{1}{\tau_1} = \left(\frac{h}{m^*\sqrt{2\pi}}\right)\left(\frac{l_e}{\sqrt{n_{2D}}}\right)^{-1} \tag{1}$$

where l_e is the mean feature size, assuming inelastic scattering occurring at the boundaries. The background impurity density has subsequently been experimentally measured at ~2 × 10^{13} cm^{-3} through low temperature Hall measurements on AlInSb flat layers. A typical graph of the modelled temperature dependant mobility, including a non-parabolic effective mass, is shown in figure 3 left, excluding surface feature scattering. Figure 3 right is modified to include surface feature scattering following equation (1). As can be seen, the model matches well the measured data for a sample with a 3 K mobility of approximately 200,000 $cm^2V^{-1}s^{-1}$. In the low temperature regime the dominant scattering mechanism is associated with the surface features, whereas at high temperature, the mobility is dominated by phonon scattering processes.

4. Modelling of surface feature potential profile

In the transport model above, it was assumed that surface features are regions of constant potential where transport occurs ballistically, whereas the boundaries between features act as potential barriers (scattering centres). These features are believed to be caused by screw-like growth around a threading dislocation [7], with the barriers at the boundaries due to the nonconformity in the crystal structure, giving rise to effective barrier widths of several monolayers. Landauer-Büttiker theory has been used to calculate tunnelling currents, and a combination of a Monte-Carlo (MC) simulation and a simple Drude model has been used to calculate corresponding mobilities. The measured and calculated parameters for the sample modelled are given in table 1, where the number of barriers is given by the length of the Hall bar divided by the average feature size. The width of the barriers is assumed negligible compared to the spacing between barriers.

Table 1. Measured and calculated parameters of sample modelled.

Parameter	Symbol	Value	Parameter	Symbol	Value
Number of barriers	N	82	Experimental current	I	1 μA
Carrier density	n_{2D}	3 × 10^{11} cm^{-2}	Experimntal voltage	V	0.4 mV
Limiting mobility	μ	270,000 $cm^2V^{-1}s^{-1}$	Mean feature size	l_e	2.43 μm

Following references [18] and [19], a tunneling current density, J, can be obtained for a confined 2D sheet of charge (xy) tunneling through barriers in the y direction for an applied voltage, V,

$$J = \frac{2e}{h}\int_0^\infty \left[n_{1D}(E_F - E_y) - n_{1D}(E_F - E_y - eV)\right]T(E_y)dE_y \tag{2}$$

$$n_{1D}(E_F) = \frac{\sqrt{2m^*}}{h}\int_0^\infty \left(\sqrt{E_x}\right)^{-1}\left[1 + \exp((E_x - E_F)/k_BT)\right]^{-1}dE_x \tag{3}$$

where E_F is the Fermi energy (calculated from n_{2D}, accounting for non-parabolic effects using a six band Kane model [3, 17, 18]), T is the temperature, k_B is the Boltzmann constant and $T(E_y)$ is the energy dependent transmission coefficient. For a specific barrier height, width and applied voltage, $T(E_y)$ can be obtained numerically following Tsu and Esaki [19] or Ando and Itoh [20]. Once the Monte Carlo simulation is performed, an effective $T^*(E_y)$ for N barriers can be calculated using the simulation results, however this is expected to be approximately equal to $T(E_y)$.

Barrier heights were determined for barriers from 1 to 50 monolayers (MLs) by calculating $T(E_y)$ and J and matching this to the values in table 1. Barrier heights (in eV) were found to approximately follow the form: height $\approx$ width$^{-1.25} \times 9.41$, where width is measured in MLs. Subsequently, the MC simulation was run to determine a corresponding mobility. For the MC model it is assumed [21] that the left contact is a perfect source/emitter/reflector and the right contact is a perfect sink/acceptor. The electrons are non-interacting and travel at a constant velocity, v, between barriers, determined by their energy. As applied voltage is small compared to barrier heights, this is neglected. Each barrier is identical and separated by the mean feature size, l_e. If a particle is transmitted, no collision occurs, whereas a refection is a scattering event. The mobility can then be calculated as an average over particles by using a simple Drude model,

$$\mu = \frac{e\tau}{m^*} = \frac{e}{m^*}\overline{\left(\frac{t}{N_c}\right)} = \frac{e}{m^*}\frac{l_e}{v} \times \overline{\left(\frac{N_s}{N_c}\right)} \quad (4)$$

where τ is the average time between collisions, t the total time taken for a particle to travel from contact to contact, N_c is the corresponding number of collisions and N_s is the number of steps taken. At finite temperatures, conduction occurs over a spread of energies proportional to the rate of change of the Fermi distribution, f, and so an energy average of mobility must be performed, accounting for a temperature dependent band gap, Fermi energy and effective mass at each energy [22].

The resultant mobilities for barriers from 1 to 50 MLs are shown in figure 4. This is a replot of figure 3, replacing the surface feature scattering given by equation (1) with that from the MC model, and as before, the total mobility closely matches that experimentally measured. The inset shows $T^*(E_y)$ at 3 K (solid lines) and Fermi distributions as a function of energy for 3 K and 300 K (dashed). The colour of the surface feature scattering and of $T^*(E_y)$ represents the barrier width in monolayers.

Figure 4 clearly shows the low temperature mobility is consistent for all barrier widths (due to the approximately constant value of $T^*(E_y)$ at $E = E_F$). At higher temperatures the mobility trends for each barrier width begin to diverge (due to the relation between $T^*(E_y)$ and the Fermi distribution), however the different barrier shapes dependencies' do not diverge until T is greater than ~70 K. At this temperature, transport modelling shows phonon scattering is dominant, and it is therefore not possible to precisely determine the shape of potential barrier due to surface features on this sample.

These figures do show however that the low temperature limiting scattering due to surface features can be modelled as a series of potential barriers, with a range of widths, matching measured currents and Hall mobilities to those calculated through tunneling currents and Monte Carlo simulations. It can also be seen from the figure that above 50 ML the surface feature mobility begins to grow exponentially at higher temperatures. It is reasonable therefore that barrier thicknesses must be smaller than 50 ML, consistent with the possible cause of screw-like growth around a threading dislocation, with micro twin defects at the boundaries.

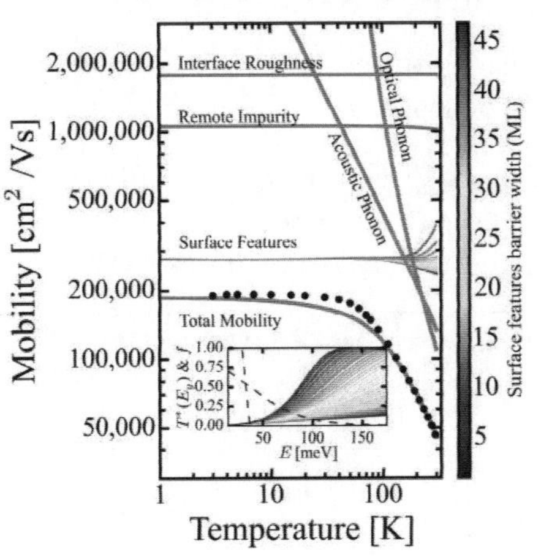

Figure 4. Reproduction of figure 3 with the MC modelled mobility due to scattering from surface features for 1 ML (bottom) to 50 ML (top) width barriers. Inset: Transmission $T^*(E_y)$ at 3 K vs energy (solid lines) for barriers from 1 ML (bottom) to 50 ML (top) as well as Fermi distribution f vs energy for 3 K and 300 K. For surface feature scattering and $T^*(E_y)$, colours represent barrier widths (see online for colour version of figure).

IOP Conf. Series: Journal of Physics: Conf. Series **964** (2018) 012005 doi:10.1088/1742-6596/964/1/012005

5. Conclusions

We have studied InSb/AlInSb QW 2DEG material and have shown evidence for three characteristic regimes in the measured low temperature Hall mobility behaviour with carrier density. We have demonstrated the application of image analysis techniques to extract representative feature sizes and through use of a transport lifetime model, shown these to be the dominant low temperature mobility limiting scattering mechanism. We have subsequently used Monte-Carlo simulations to model the current, and the limiting mobility due to these features. Through use of a combination of Landauer-Büttiker and Drude models, we have demonstrated the possible range of potential barrier heights and widths, from 1 ML to 50 ML, and shown these to be consistent with scattering corresponding to surface features as the dominant scattering mechanism. This work shows that with correct buffer redesign there is a clear potential for significant improvement in the mobility of such material.

Acknowledgements

The work was supported by the UK Engineering and Physical Sciences Research Council [grant numbers EP/L012995/1 and EP/M507842]. Data supporting this research is openly available from the Cardiff University Research Portal at http://dx.doi.org/10.17035/d.2017.0040827720.

References

[1] Madelung O, Rössler U and Schulz M 2002 *Landolt and Bornstein - Group III Condensed matter: Group IV Elements, (IV-IV) and (III-V) Compounds. Part b - Electronic, Transport, Optical and Other Properties* (Berlin: Springer).
[2] Yi W *et al* 2015 *Appl. Phys. Lett.* **106** 142103
[3] Pooley O J, Gilbertson A M, Buckle P D, Hall R S, Emeny M T, Fearn M, Halsall M P, Cohen L F and Ashley T 2010 *Semicond. Sci. Technol.* **25** 125005
[4] Orr J M S, Gilbertson A M, Fearn M, Croad O W, Storey C J, Buckle L, Emeny M T, Buckle P D and Ashley T 2008 *Phys. Rev.* B **77** 165334
[5] Qu F *et al* 2016 *Nano Lett.* **16** 7509-13
[6] Zhao X, Zhang Y, Guan M, Cui L, Wang B, Zhu Z and Zeng Y 2017 *J. Cryst. Growth* **470** 1-7
[7] Shi Y, Gosselink D, Gharavi K, Baugh J and Wasilewski Z R 2017 *J. Cryst. Growth* (*Preprint*)
[8] McIndo C J, Hayes D G, Papageorgiou, Hanks L A, Smith G V, Allford C P, Zhang S, Clarke E M and Buckle P D 2017 *Physica* E **91** 169–72
[9] Hayes D G, Allford C P, Smith G V, McIndo C, Hanks A L, Gilbertson A M, Cohen L F, Zhang S, Clarke E M and Buckle P D 2017 *Semicond. Sci. Technol.* **32** 085002
[10] Mourik V, Zuo K, Frolov S M, Plissard S R, Bakkers E P and Kouwenhoven L P 2012 *Science* **336** 1003–7
[11] Deng M T, Yu C L, Huang G Y, Larsson M, Caroff P and Xu H Q 2012 *Nano Lett.* **12** 6414–9
[12] Nedniyom B Nicholas R J, Emeny M T, Buckle L, Gilbertson A M, Buckle P D and Ashley T 2009 *Phys. Rev.* B **80** 125328
[13] van den Berg J W G, Nadj-Perge S, Pribiag V S, Plissard S R, Bakkers E P A M, Frolov S M and Kouwenhoven L P 2013 *Phys. Rev. Lett.* **110** 066806
[14] Li R, You J Q, Sun C P and Nori F 2013 *Phys. Rev. Lett.* **111** 086805
[15] Awschalom D D, Basset L C, Dzurak A S, Hu E L and Petta J R 2013 *Science* **339** 1174–9
[16] Tan I-H, Snider G L, Chang L D and Hu E L 1990 *J. Appl. Phys.* **68** 4071
[17] Gilbertson A M, Branford W R, Fearn M, Buckle L, Buckle P D, Ashley T and Cohen L F 2009 *Phys. Rev.* B **79** 235333
[18] Davies J H 1998 *The Physics of Low-Dimensional Semiconductors: An Introduction* (Cambridge: Cambridge University Press)
[19] Tsu R and Esaki L 1973 *Appl. Phys. Lett.* **22** 562
[20] Ando Y and Itoh T 1987 *J. Appl. Phys.* **61** 1497
[21] Bagwell P F and Orando T P 1989 *Phys. Rev.* B **40** 1456
[22] Asgari A and Faraone L 2011 *J. Appl. Phys.* **110** 113713

IOP Conf. Series: Journal of Physics: Conf. Series **964** (2018) 012006 doi:10.1088/1742-6596/964/1/012006

Transport of modulation-doped $Al_{0.2}Ga_{0.8}Sb$/GaSb heterojunctions

L A Hanks[1], M Hayne[1], A R J Marshall[1] and L Ponomarenko[1]

[1]Department of Physics, Lancaster University, Lancaster LA1 4YB, United Kingdom

Email: l.hanks@lancaster.ac.uk

Abstract. Mobilities and carrier densities of modulation doped $Al_{0.2}Ga_{0.8}Sb$/GaSb heterostructures are presented for the first time. The structures studied were grown by molecular beam epitaxy and consisted of a single heterojunction with Te compensation doping to reduce the intrinsic p-type background. Hall measurements were performed from 30–300 K, giving p-type mobilities peaking at 3240 cm^2/Vs, a considerable improvement over previous reported bulk mobilities for samples with compensation doping. Growth trials on bulk material have also been carried out to investigate the optimum growth conditions for future structures, with the aim of minimising the occurrence of natural growth defects in GaSb, which act as acceptors. Together these measurements lay the ground work for (magneto)transport studies of two-dimensional charge-carriers in $Al_xGa_{1-x}Sb$/GaSb heterostructures, which has not been previously reported.

1. Introduction

GaSb has been known and well-studied as an optoelectronic material for many years, but knowledge of its electronic properties is limited [1]. It has an electron effective mass of 0.039 m_e (where m_e is the mass of the electron), 0.58 times that of GaAs, and a narrow band gap of 0.812 eV at 0 K. In many ways, GaSb is a close relation of GaAs, with minimal lattice mismatch to its natural tertiary barrier material (AlGaSb) [2]. However, development in GaAs transport structures have far surpassed GaSb, which has been limited to basic bulk measurements, with this stagnating in the 1990's. A prime reason for the vast knowledge on GaAs is the availability and cheap cost of high-quality substrates for subsequent growth, from which various confined systems have been created, giving us a wealth of understanding of the material. Other III-V semiconductors, such as GaSb, have to compromise their material quality when grown on a GaAs substrate due to the lattice mismatch. Here, to decrease the strain we incorporate an interfacial misfit array (IMF) layer which significantly reduces threading dislocation in the growth direction (measurement plane), leading to higher quality material more suited to electrical applications [3]. Further improvements can be gained from modulation doping and confinement which have not yet been realised in GaSb for transport measurements.

Hole transport investigations in GaSb are much more extensive than electron transport, with various groups having investigated the low-temperature range and multiple carrier concentrations. These are generally on unintentionally p-doped or insufficiently compensation n-doped bulk GaSb structures. GaSb is a nominally p-type material with a high acceptor defect producing a carrier concentration of $\sim 1\times10^{17}$ cm^{-3} [4]. From various investigations, the cause of this high concentration has been found to be a native defect in the form of a doubly accepting gallium antisite (Ga_{Sb}) [5]. Previous growth studies have shown that by lowering the growth temperature and the V/III flux ratio, these intrinsic defects can

IOP Conf. Series: Journal of Physics: Conf. Series **964** (2018) 012006 doi:10.1088/1742-6596/964/1/012006

be decreased to 1×10^{16} cm^{-3}, and by changing the growth further, n-type GaSb is achievable through a different defect, though the transport details of this are not known [6].

Clearly, there is room for expansion in the field of electron and hole transport in GaSb. Our work investigates transport in $Al_{0.2}Ga_{0.8}Sb$/ GaSb heterojunctions, with Te modulation doping to compensate the p-type background, avoiding the creation of more scattering centers in the conducting channel. This lays the groundwork for creation and (magneto)transport measurement of two-dimensional charge carriers in $Al_xGa_{1-x}Sb$/GaSb heterostructures.

2. Growth, fabrication and experimental details

All structures were grown using molecular beam epitaxy (MBE) on semi-insulting GaAs substrates with an IMF layer at the GaSb/GaAs interface to prevent threading dislocations. Two different sets of samples were grown: $Al_{0.2}Ga_{0.8}Sb$/GaSb heterostructures and bulk GaSb.

For the heterostructures 1 μm of GaSb was grown on GaAs, followed by a 20 nm spacer layer of undoped $Al_{0.2}Ga_{0.8}Sb$ and then 100 nm of n-type $Al_{0.2}Ga_{0.8}Sb$ (Te doped). These modulation-doped samples were grown with a substrate temperature of 505 °C, a Sb/Ga growth rate ratio of 2.2, these are the standard growth conditions used in this study (STSR) and a Ga_2Te_3 source provided the doping. Both slab and δ-plane doping were achieved. The structure is shown in the inset of figure 1, where the δ-plane equivalent structure has the same layer dimensions and spacer as the slab, and a sheet concentration of 1×10^{12} cm^{-2}. From bulk trials the intrinsic hole concentration for these growth conditions is estimated to be ~5×10^{17} cm^{-3}. The $Al_{0.2}Ga_{0.8}Sb$ is doped enough to partially compensate, but not form a triangular well at the junction, therefore transport in the thick GaSb layer (1 μm) is considered bulk. Hall bars with an aspect ratio of ~30:1 were created using standard top-down photolithography fabrication techniques. Ohmic contacts were achieved with a multilayer metal composition of Pd/In/Pd/Au (5/40/4/100 nm), however, they were high resistance. Low-frequency DC Hall measurements were performed over a range of temperatures from 30–300K in magnetic fields up to ±0.5 T.

Growth trials were carried out on bulk structures of 2 μm of GaSb on GaAs, grown at various temperatures and V/III ratios, giving unintentional hole concentrations that varied between 4 and 20 $\times10^{16}$ cm^{-3} at room temperature. Further details of the growth of these structures can be found in section 4. Tellurium doping was varied and gave rise to electron concentrations up to 6×10^{17} cm^{-3}. Unlike the modulation doped structures, contact was made using indium solder in a ~ 1 cm^2 van der Pauw geometry.

3. Mobility and carrier concentration in $Al_{0.2}Ga_{0.8}Sb$/ GaSb heterojunctions

The transport properties of the modulation-doped samples are shown in figures 1 and 2. As there is minimal difference between the slab and delta samples for both the carrier concentration and the mobility, only the slab doped sample is shown in figure 2. It can be seen in figure 2 that carrier density decreases exponentially with decreasing temperature, with the onset of freeze out at ~150 K. The mobility steadily increases with decreasing temperature until 70 K, where it reaches a maximum before sharply decreasing at low temperature as the carriers freeze out.

A two carrier fit was implemented on the conductivity data of all measurements, following the method of Reed *et al.*[7]. From this analysis it was concluded that the samples can be considered single carrier as a significant majority of the carriers across all temperatures are holes. Hole concentration peaked at 2.2×10^{16} cm^{-3}, with the electron density consistently being calculated as <6% of the total carrier concentration.

IOP Conf. Series: Journal of Physics: Conf. Series **964** (2018) 012006 doi:10.1088/1742-6596/964/1/012006

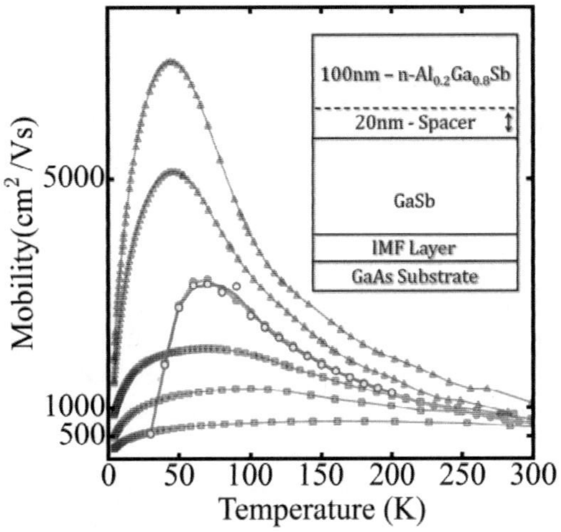

Figure 1. Mobility against temperature for modulation doped samples, slab doped (○) and delta doped (●) compared to bulk structures grown and measured by Dutta *et al.* [8] Te compensated (□) and pure GaSb (▲).

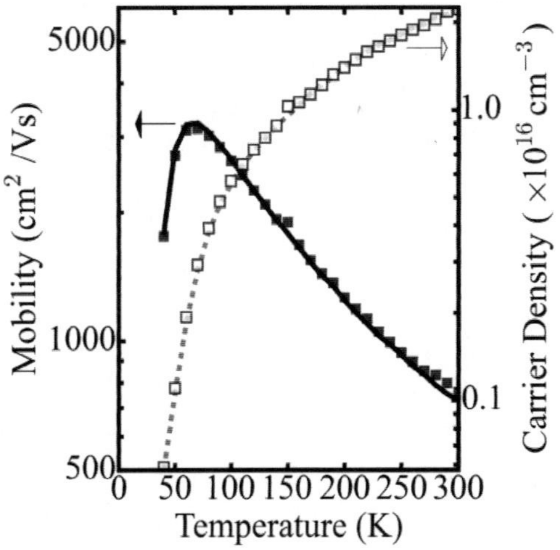

Figure 2. Mobility as a function of temperature for a GaSb slab doped heterojunction (-), calculated hole mobility (■) with their associated carrier densities (--,□).

Previous work investigating scattering mechanisms and defects in GaSb was undertaken by Dutta *et al* on bulk GaSb structures, both undoped and compensation-doped with Te. These structures were entirely GaSb single crystals, and so had no strain defects [8]. Their work concluded that in the undoped samples the mobility was limited by phonons above 45 K and by ionized impurities below 45 K, whereas the doped samples were ionized impurity limited at all temperatures. By comparing to their work on Te compensated samples (squares in figure. 1), it is shown that our samples have higher mobility between 40 and 200 K, despite having more potential strain, and form a curvature more similar to the pure samples than the doped. The carriers in our sample close to the junction will be scattered by the Te remote ionized impurities, but will this have only a small effect on the majority in the thick GaSb conducting channel. Therefore, the extrinsic impurity scattering in our samples is minimal compared to the scattering created by Te ionized impurities in the bulk doped samples studied by Dutta *et al.* Above 200 K the mobilities are comparable, and below 70 K carrier freeze-out causes a decrease in our mobility.

Overall, our results show good promise for future high-mobility (magneto)transport measurements on $Al_xGa_{1-x}Sb$/GaSb heterostructures. Further to these samples, a growth trial has been carried out to explore the defect density's dependence on growth conditions, with the aim of obtaining even higher mobility material.

4. Growth study of bulk GaSb

It has been known for many years that GaSb's high acceptor concentration primarily originates from defects in the growth in the form of a gallium antisite (Ga_{Sb}); a gallium atom in the lattice site of an antimony atom [5]. These defects occur regardless of the growth method but are heavily dependent on growth conditions. Due to the less than unity sticking coefficient of Sb, a V/III growth ratio of more than 1 is required at all times. However in the case of GaSb, counter intuitively, a greater Sb excess leads to an increase in GaSb defects, creating a very narrow optimum growth window for GaSb with low unintentional doping. To investigate this relationship between growth conditions and defect density, we have performed growth trials on MBE-grown GaSb samples. Figure 3(a) shows the variation in measured carrier density with intended doping carrier concentration. The intended doping concentration is determined from a series of InSb calibration samples, with appropriate correction for the growth rate of GaSb.

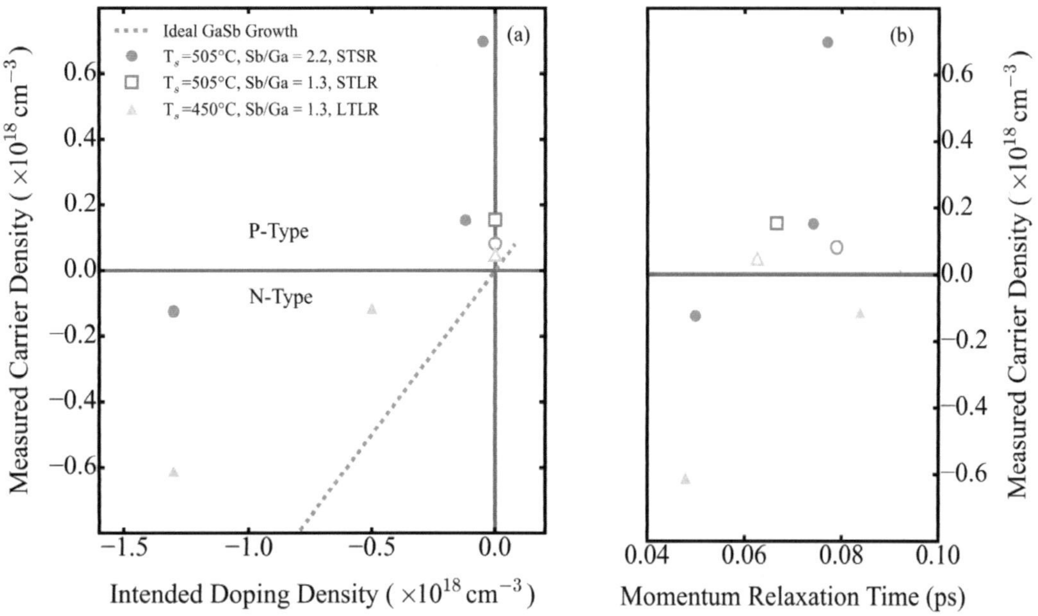

Figure 3. (a) Measured carrier density against intended doping density at room temperature for 3 growth conditions (key shown in the figure). Undoped samples are shown as open symbols and the ideal GaSb growth, shown as a dashed line, is a 1:1 relationship between intended doping density and measured carrier density. **(b)** Momentum relaxation time against measured carrier density. The positive measured carrier density represents hole concentration and the negative carrier density represents electron concentration. Symbols have the same meaning as in (a).

It can be seen from figure 3(a) that for undoped samples (zero on horizontal axis), a lower growth temperature and V/III growth ratio (triangle) yields the most reduced defect density and is named the LTLR growth condition. For increased intended n-type doping, measured carrier density does not follow the ideal 1:1 ratio, veering away from the ideal case. Despite this, a reasonable n-type carrier concentration of 6×10^{17} cm^{-3} is obtained with the LTLR growth condition, so these conditions give a measured carrier density that is closest to the ideal at all intended doping densities.

Figure 3(b) shows the corresponding room temperature momentum relaxation time for each of the samples in figure 3(a). For the undoped samples (open symbols) the longest momentum relaxation time is for the growth conditions of STSR growth condition (circle). However, the difference is modest. In the p-type regime the momentum relaxation time for this growth condition appears approximately invariant with measured carrier density. This may suggest that for p-type samples mobility is phonon limited at room temperature.

In the n-type regime, it can be seen that the LTLR samples show a decrease in momentum relaxation time with increasing n-type carrier density, suggesting, much like Dutta *et al.*, a greater influence from ionized impurity scattering. The equivalent STSR n-type sample (circle) has a reduced momentum relaxation time compared with the LTLR sample of the same carrier density, suggesting that in the n-type regime a LTLR growth produces less scatter and therefore has a better material quality. To add certainty to these results a more varied growth trial would need to be carried out.

Overall, this points to a lower temperature and V/III flux ratio giving improved growth, resulting in improved carrier densities and mobilities. Thus, the results of this growth trial allow for higher mobility material for future (magneto)transport measurements on $Al_xGa_{1-x}Sb$/GaSb heterostructures.

5. Conclusions

We have reported, for the first time, an investigation of the transport properties of modulation-doped $Al_xGa_{1-x}Sb$/GaSb heterojunctions, and growth trials on bulk GaSb. The mobility of modulation compensation-doped samples was greater than for bulk compensation-doped samples due to the reduced scattering from remote ionized impurities compared to bulk dopants. The mobility of the slab modulation-doped sample peaked at 3240 cm^2/Vs at 70 K for a carrier concentration of 2.8×10^{17} cm^{-3}, with similar values for the δ doped sample. A bulk growth study concluded that low growth temperature and low V/III reduces the number of defects, which create the high p-type background in GaSb. This improves the momentum relaxation time for n-type GaSb, and should have a similarly beneficial effect in modulation-doped samples. Thus, this work sets the scene for (magneto)transport studies of high-mobility two-dimensional charge carriers in $Al_xGa_{1-x}Sb$/GaSb heterostructures.

Acknowledgments

This work was supported by the Engineering and Physical Sciences Research Council. The figures used in this work is have publically available datasets available at DOI https://dx.doi.org/10.17635/lancaster/researchdata/173

References

[1] Dutta P S, Bhat H L and Kumar V 1997 *J. Appl. Phys* **81** 5821-70
[2] Vurgaftman I, Meyer J R and Ram-Mohan L R 2001 *Appl. Phys. Rev.* **89** 5815-75
[3] Craig A P, Carrington P J, Liu H, and Marshall A R J 2016 *J. Cryst. Growth* **435** 56-61
[4] Lee H J and Wolly J C 1981 *Can. J. Phys.* **59** 1844-50
[5] Virkkala V, Havu V,Tuomisto F and Puska M J 2012 *Phys. Rev.* B **86** 144101
[6] Chen S M and Su Y K 1993 *J. Appl. Phys.***74** 2892-95
[7] Reed M A, Kirk W P and Kobiela P S 1986 *IEEE J. Quantum Electron.* **QE-22** 1753-59
[8] Dutta P S, Prasas V, Bhat H L and Kumar V 1996 *J. Appl. Phys* **80** 2847-53

IOP Conf. Series: Journal of Physics: Conf. Series **964** (2018) 012007 doi:10.1088/1742-6596/964/1/012007

Landauer current and mutual information in a bosonic quantum dot

Hrushikesh Shashikant Sable
Physical Research Laboratory, Ahmedabad, 380009, India
E-mail: hsable@prl.res.in

Devendra Singh Bhakuni
Indian Institute of Science Education and Research Bhopal, M.P, 462066, India
E-mail: devendra123@iiserb.ac.in

Auditya Sharma
Indian Institute of Science Education and Research Bhopal, M.P, 462066, India
E-mail: auditya@iiserb.ac.in

Abstract. We study the quantum transport of bosons through a quantum dot coupled to two macroscopic heat baths L and R, held at fixed temperatures T_L and T_R respectively. We manage to cast the particle as well as the heat current into the Landauer form. Following the correlation matrix approach, we compute the time-dependent mutual information of the dot with the baths. We find that mutual information goes logarithmically as the number of bosons, and at low temperatures, it is possible to set up the parameters in such a way that in steady-state, the mutual information goes quadratically as a function of current.

1. Introduction

Quantum entanglement [1] has proven to be a useful quantity to probe in a variety of phenomena. The last decade or so has seen a great proliferation of activity at the interface between quantum information and condensed matter physics [2, 3], two disciplines that have been traditionally considered to be distinct and rather far apart. One specific context in which this merger has shown itself to be particularly interesting, is that of quantum transport. Although the Landauer approach to quantum transport [4] [5] [6] has been around for many decades, and has been developed extensively, only recently [7] has the connection between current through a quantum dot, and the quantum correlations that develop between the dot and the leads in a nonequilibrium setting, been shown to be intimate. A motivation for the current study is to explore how this connection plays out when the statistics involved is bosonic, rather fermionic. When bosonic degrees of freedom are involved, in addition to the particle current, a *heat current* also becomes relevant [8, 9, 10]. The drive for the dynamics is generated by a temperature gradient between the baths.

When quadratic Hamiltonians are involved, Wick's theorem can be exploited to provide a prescription for computing entanglement in the eigenstates in terms of the eigenvalues of an

IOP Conf. Series: Journal of Physics: Conf. Series **964** (2018) 012007 doi:10.1088/1742-6596/964/1/012007

underlying correlation matrix [11, 12]; this approach has been successfully employed in a variety of contexts. However, the literature on entanglement in nonequilibrium phenomena [13], and particularly for bosonic sytems [14], is relatively sparse. A nonequilibrium system typically involves mixed states, where entanglement is rather hard to study. Mutual information, [15, 16, 17] which includes both classical and quantum correlations is one of the commonly studied quantities in such a scenario, and we too adopt this for our study.

The layout of the current paper is as follows. In the next section, we describe our model, and the Ohmic bath spectral density used, from which the parameters are obtained. In the following section, we write down the exact non-equilibrium density matrices for the dot and the baths. Introducing a new set of bosonic operators, we calculate the current in the subsequent section. We learn that expressions for the particle current calculated and heat current calculated are very similar, and can be recast in the Landauer form. Next, exploiting the fact that the reduced density matrices have a thermal form at all times in the dynamics, the spectra of the reduced density matrices can be related to the time-dependent correlation matrices. We work out the expression for the von Neumann entropy of the subsystems and calculate the mutual information. A results and discussion section then collects and compares current and mutual information. The last section is reserved for conclusions and summary.

2. Model Hamiltonian

The model consists of a quantum dot in the center coupled to two bosonic baths - left L and right R. The Hamiltonian can be written as

$$\begin{aligned} H &= H_L + H_R + H_D + H_{LD} + H_{RD} \\ H_D &= \epsilon_d d^\dagger d \\ H_{L,(R)} &= \sum_{k\in L(R)} \epsilon_k c_k^\dagger c_k \\ H_{LD,(RD)} &= \sum_{k\in L(R)} t_k (c_k^\dagger d + c_k d^\dagger), \end{aligned} \tag{1}$$

where ϵ_d is energy of the dot, ϵ_k is the kth mode energy of a bath and t_k is the coupling of the dot and the kth mode. Throughout the paper we set $\hbar = 1, e = 1$ and $k_B = 1$. One can obtain the couplings t_k from the spectral density function $J(\omega)$, for which a general expression is [18]

$$J(\omega) = \sum_k^N t_k^2 \delta(\omega - \omega_k). \tag{2}$$

Following the standard approach for bosons [19], we model the baths using the Ohmic spectral density with an exponential cut-off ω_c at both ends (L, R) [20]:

$$J(\omega) = \eta\omega e^{\frac{-\omega}{\omega_c}}, \tag{3}$$

where η is the damping/friction constant. Integrating in a small $\Delta\omega$ window around ω_k, we obtain

$$t_k = \sqrt{\eta\omega_k e^{\frac{-\omega_k}{\omega_c}} \Delta\omega}. \tag{4}$$

3. Non-equilibrium density matrix

Initially the left and right baths are separately in equilibrium, each at zero chemical potential and different temperatures T_L and T_R respectively. So the initial state of the system can be described by

$$\rho(0) = \rho_L(0) \otimes \rho_R(0) \otimes \rho_D(0), \tag{5}$$

where the left and the right baths are in thermal equilibrium:

$$\rho_L(0) = \frac{\exp(-\beta_L \hat{H}_L)}{Z_L}, \quad \rho_R(0) = \frac{\exp(-\beta_R \hat{H}_R)}{Z_R}, \tag{6}$$

and the dot's state is represented as $\rho_D(0) = n_0 d^\dagger d + (1 - n_0) d d^\dagger$. At $t = 0$, the coupling Hamiltonian $H_{LD,(RD)}$ is turned on and the unitary time evolution of the density matrix $\rho(t) = e^{-iHt}\rho(0)e^{iHt}$ is governed by the Hamiltonian in Eq. 1. The Hamiltonian becomes diagonal when expressed in terms of new bosonic operators $a_\alpha = \sum_{i=1}^{N} \psi_\alpha(i) c_i$, where $\psi(i)$ is the ith eigenvector of H corresponding to eigenvalue e_i [7], where N is the total number of levels including the baths and the dot, $N = N_L + N_R + 1$, N_L and N_R being the energy-levels in the left and right baths respectively. The $N^{\rm th}$ index is used for the dot.

4. Current in Landauer form

The left(right) particle current is the rate of change of occupancy of left (right) bath.

$$I_{L/R}(t) = -\frac{d\langle \hat{N}_{L/R} \rangle}{dt} = i \sum_{i \in L(R)} t_i \big\langle (c_i(t) c_N^\dagger(t) - c_i^\dagger(t) c_N(t)) \big\rangle. \tag{7}$$

The overall current is defined as $I(t) = \frac{I_L(t) - I_R(t)}{2}$ which simplifies to [7]

$$I(t) = \sum_k t_k' \sum_{i,j,p=1}^{N} \text{Im}\Big(\psi_i(N)\psi_j^*(k)\psi_i^*(p)\psi_j(p)e^{i(e_i - e_j)t}\Big) f_p, \tag{8}$$

where $t_k' = t_k$ if $k \in L$ and $t_k' = -t_k$ if $k \in R$ and f_p is the Bose Einstein function $f(\epsilon_p, T_L)$ for modes on the left bath, $f(\epsilon_p, T_R)$ for modes on the right bath and n_0 for the dot population. The Landauer form of the current is [21]

$$I(E) = \int \mathcal{T}(E)\Big(f_L(E) - f_R(E)\Big)dE, \tag{9}$$

where the $\mathcal{T}(E)$ is the transmission function and f_L and f_R are the distribution functions for the left and the right bath respectively. The symmetry present in the energy levels and the coupling constants of the baths allows us to express the current in the Landauer form as

$$I(t) = 2\sum_{p=1}^{N_L}\sum_{k=1}^{N_L} t_k \sum_{i=1,3,.;j=2,4.}^{N} \text{Im}\Big(\psi_i(N)\psi_j^*(k)\psi_i^*(p)\psi_j(p)e^{i(e_i - e_j)t}\Big)\Big(f_B(T_L) - f_B(T_R)\Big). \tag{10}$$

In analogy with the particle current, one can work out the expression for the heat current as

$$J_{L/R}(t) = -\frac{d\langle \hat{H}_{L/R} \rangle}{dt} = i \sum_{i \in L(R)} t_i \epsilon_i \big\langle (c_i(t) c_N^\dagger(t) - c_i^\dagger(t) c_N(t)) \big\rangle. \tag{11}$$

The expression for heat current has the same form as the particle current:

$$J(t) = \sum_k t_k' \epsilon_k \sum_{i,j,p=1}^{N} \text{Im}\Big(\psi_i(N)\psi_j^*(k)\psi_i^*(p)\psi_j(p)e^{i(e_i - e_j)t}\Big) f_p. \tag{12}$$

One can also recast the heat current into Landauer form as in Eq.10 with t_k replaced by $t_k \epsilon_k$.

IOP Conf. Series: Journal of Physics: Conf. Series **964** (2018) 012007 doi:10.1088/1742-6596/964/1/012007

5. Mutual information

We wish to study the evolution of the quantum correlations between the quantum dot and the baths. Since the initial state is mixed, von Neumann entropy is unsuitable as a measure of entanglement. Therefore we study mutual information which measures the total correlations: quantum and classical. The mutual information between the dot and the baths is defined as

$$S = S_D + S_{LR} - S_{\text{full}}, \tag{13}$$

where $S_D = -\text{Tr}(\rho_D \ln \rho_D)$, $S_{LR} = -\text{Tr}(\rho_{LR} \ln \rho_{LR})$ and $S_{\text{full}} = -\text{Tr}(\rho \ln \rho)$ corresponds to the von Neumann entropies of the dot, baths and the full system respectively. It can be shown that the reduced density matrices have a thermal-like form at all times in the dynamics [7]. This fact allows for the characterization of the reduced density matrix of a subsystem by the single particle correlators only [12]. Then the von Neumann entropy S_G of a subspace G is given as [11, 22]

$$S_G = \sum_{\sigma=1}^{N_G} [(1 + C_\sigma)\ln(1 + C_\sigma) - C_\sigma \ln(C_\sigma)], \tag{14}$$

where C_σ are the eigenvalues of the correlation matrix defined within subspace G and N_G is the total number of sites in the subspace. The above formula is also valid for a time dependent correlation matrix. Thus to calculate mutual information, we need to calculate the correlation matrix of the baths and the dot at each time step and diagonalize it to obtain the eigenvalues.

6. Results and discussion

We now present numerical results for a large but finite system with N_L=128, N_R=128 sites in the left and the right bath respectively. We have verified that this system size is large enough to have converged to the thermodynamic limit (by the indistinguishability of data obtained for all the quantities from a system of half the size). To generate the spectral density defined in Eqn. 3, we choose the cut-off $\omega_c = 20$ and discretization $\Delta \equiv \omega_c/N_L$. The energy levels in each of the baths range uniformly starting from a small positive value of Δ upto ω_c in steps of Δ.

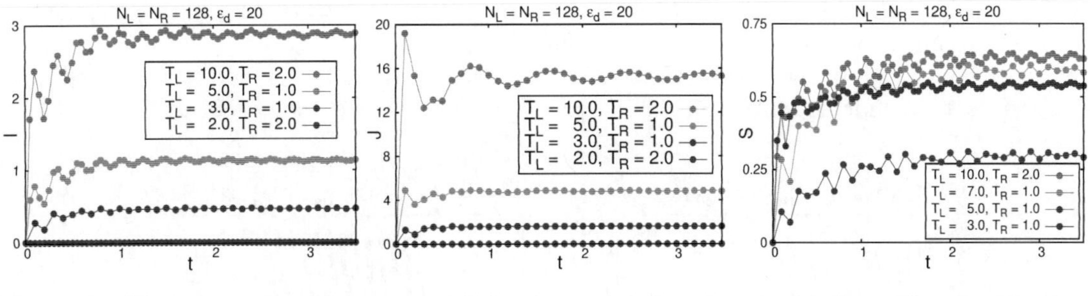

Figure 1: Variation of particle current I, heat current J and mutual information S with time. A nonequilibrium steady state is attained at long times.

The following quantities: particle current I, heat current J and mutual information S are plotted in Fig. 1 as a function of time. We see a close relationship between I and S as well as J and S. Greater the current is, greater are the correlations between the baths and the dot. Thus non-equilibrium current works as a physical observable to measure the correlations in the system [7]. Since the transient behaviour of the current and the mutual information involves rapid changes, we focus on their steady state values. We define a dimensionless ratio $r = \frac{T_L - T_R}{T_L + T_R}$ and plot the steady state value of the current I_∞ and the mutual information S_∞ for different ratios r.

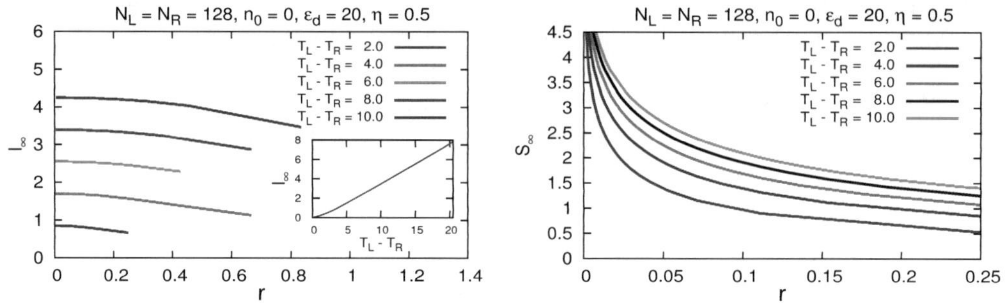

Figure 2: The steady state value of current and mutual information plotted against the ratio r for various temperature differences. The steady state value of current only depends on the temperature differences as depicted by the inset. S_∞ decreases on increasing r, or increases on increasing both T_L and T_R.

From Fig. 2, we see that the current depends only on the temperature difference $T_L - T_R$ at high temperatures, which is a direct consequence of the Landauer formula. We also observe that S_∞ increases on increasing both T_L and T_R. Higher temperatures T_L and T_R correspond to a

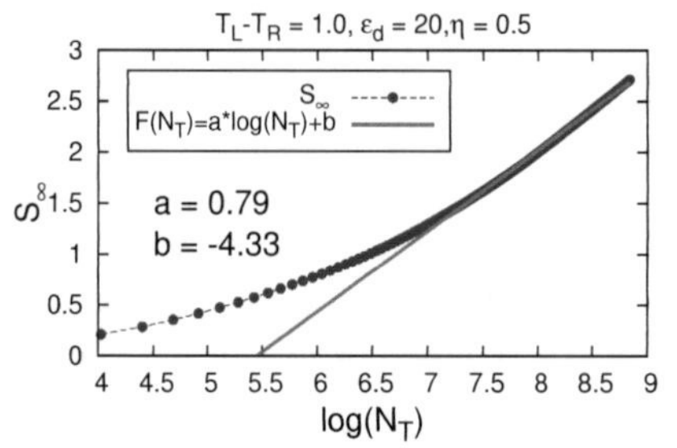

Figure 3: A logarithmic behaviour of mutual information with the total number of bosons N_T is obtained for a large N_T. The total number of bosons N_T can be calculated from the distribution function as $N_T = \sum_{i=1}^{N_L} f_B(\epsilon_i, \mu, T_L) + \sum_{i=1}^{N_R} f_B(\epsilon_i, \mu, T_R) + n_0$.

larger total number of bosons N_T in the left and the right baths respectively and hence greater the correlations between the baths and the dot. Fig. 3 shows that S_∞ is found to be logarithmic with respect to N_T for large N_T. Such behaviour has been reported before [22].

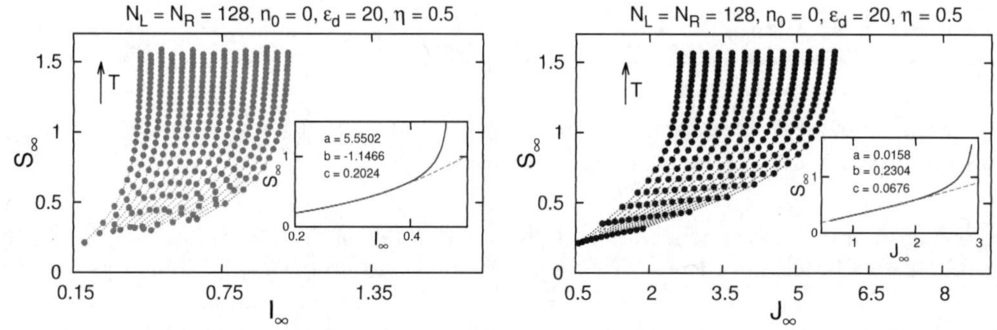

Figure 4: The relationship between the steady state value of mutual information and current (particle I_∞ and heat J_∞). Inset shows the fitting of one curve to quadratic form $ax^2 + bx + c$. Average temperature $T = \frac{T_L + T_R}{2}$ increases along the pointed direction.

We now proceed to study the behaviour of S_∞ with respect to I_∞ and J_∞. Fig. 4 are obtained by noting the steady state values of the current (particle and heat) and the mutual information for different temperature differences. Each curve corresponds to a particular ΔT. When the temperature is low and the current depends explicitly on the temperature, we see by the fit that the S_∞ goes quadratically as a function of I_∞/J_∞. In the limit of large temperatures, the current saturates to the highest possible value but the mutual information keeps on increasing. Here we have shown data for just one representative sample of parameters $\epsilon_d = 20, \eta = 0.5$, where a clean quadratic relationship between mutual information and current is obtained.

7. Summary and conclusions

We studied transport through a bosonic dot as well as the dynamics of correlations between the dot and the baths by means of mutual information. Exploiting the symmetries present in the bath energies and the couplings, we were able to cast the expression for current in Landauer form. Analogously with the fermionic version [7], the similarities in the current and the mutual information motivated us to further study the relationship between them. We found that with a suitable choice of the parameters, the steady state mutual information S_∞ depends quadratically on the steady state current I_∞ and J_∞, particularly when the initial temperatures T_L and T_R are low. We also found that this steady state value of mutual information varies logarithmically with the total number of bosons present in the system. Our findings here are for a simple noninteracting bosonic model, and it would be interesting to investigate how general these relationships can be. We are currently in the process of studying a more "realistic" Hamiltonian of bosons, that has position-position coupling giving rise to pairing terms. Also in progress is work studying other parameter regimes, where oscillatory steady states are found, and where the relationship between current and mutual information is less systematic.

Acknowledgement

We thank Eran Rabani for helpful discussions. A.S is grateful to SERB for the startup grant (File Number: YSS/2015/001696).

References

[1] Mintert F, Viviescas C and Buchleitner A 2009 *Entanglement and Decoherence* (Springer) pp 61–86
[2] Amico L, Fazio R, Osterloh A and Vedral V 2008 *Rev. Mod. Phys.* **80** 517
[3] Laflorencie N 2016 *Physics Reports* **646** 1–59
[4] Haug H and Jauho A P 2008 *Quantum kinetics in transport and optics of semiconductors* vol 2 (Springer)
[5] Landauer R 1970 *Philos Mag* **863** 21
[6] Büttiker M 1986 *Phys. Rev. Lett.* **57**(14) 1761–1764
[7] Sharma A and Rabani E 2015 *Phys. Rev. B* **91**(8) 085121
[8] Dhar A and Roy D 2006 *Journal of Statistical Physics* **125** 801–820
[9] Segal D, Nitzan A and Hänggi P 2003 *The Journal of Chemical Physics* **119** 6840–6855
[10] Taylor E and Segal D 2015 *Phys. Rev. Lett.* **114**(22) 220401
[11] Peschel I 2012 *Brazilian Journal of Physics* **42** 267–291
[12] Peschel I 2003 *Journal of Physics A: Mathematical and General* **36** L205–L208
[13] Eisler V and Zimborás Z 2014 *Phys. Rev. A* **89** 032321
[14] Eisler V and Zimborás Z 2014 *New Journal of Physics* **16** 123020
[15] Wilms J, Troyer M and Verstraete F 2011 *J. Stat. Mech.* **2011** P10011
[16] Wolf M M, Verstraete F, Hastings M B and Cirac J I 2008 *Phys. Rev. Lett.* **100** 070502
[17] De Tomasi G, Bera S, Bardarson J H and Pollmann F 2017 *Phys. Rev. Lett.* **118** 016804
[18] Breuer H P and Petruccione F 2002 *The theory of open quantum systems* (Oxford University Press)
[19] Leggett A J, Chakravarty S, Dorsey A T, Fisher M P A, Garg A and Zwerger W 1987 *Rev. Mod. Phys.* **59**(1) 1–85
[20] Weiss U 2012 *Quantum dissipative systems* vol 13 (World scientific)
[21] Meir Y and Wingreen N S 1992 *Phys. Rev. Lett.* **68**(16) 2512–2515
[22] Ding W and Yang K 2009 *Phys. Rev. A* **80** 012329

IOP Conf. Series: Journal of Physics: Conf. Series **964** (2018) 012008 doi:10.1088/1742-6596/964/1/012008

The observation of the Aharonov-Bohm effect in suspended semiconductor ring interferometers

D A Pokhabov[1,2], A G Pogosov[1,2], A A Shevyrin[1,2], E Yu Zhdanov[1,2], A K Bakarov[1,2], A A Shklyaev[1,2], S V Ishutkin[3], M V Stepanenko[3] and E V Shesterikov[3]

[1] Institute of Semiconductor Physics, Novosibirsk 630090, Russia.
[2] Novosibirsk State University, Novosibirsk 630090, Russia.
[3] Tomsk State University of Control Systems and Radioelectronics, Tomsk 634050, Russia.

E-mail: pokhabov@isp.nsc.ru

Abstract. A suspended semiconductor quantum ring interferometer based on a GaAs/AlGaAs heterostructure with a two-dimensional electron gas (2DEG) is created and experimentally studied. The electron interference in suspended 2DEG is observed. The interference manifests itself as the Aharonov-Bohm oscillations of the interferometer magnetoresistance, clearly observed before as well as after suspension. The amplitude of the oscillations remains almost unchanged after suspension.

1. Introduction

The main interest in semiconductor nanostructures with a two-dimensional electron gas (2DEG) is associated with a variety of mesoscopic phenomena observed in them, including various ballistic effects, interference of electron waves, electron-electron and spin-orbit interaction et al. They have been investigated for a relatively long time [1]. Typical semiconductor devices in which the electron interference plays a key role are the ring interferometers. Usually such devices are embedded in a semiconductor bulk [2-5]. The differing from them suspended nanostructures with a 2DEG separated from a semiconductor substrate are of particular interest. They are created by etching the sacrificial layer from under nanostructures. Electronic transport in such suspended nanostructures has features originating from additional mechanical degrees of freedom [6-15], poor heat coupling to the bulk [16-17], and electron-electron interaction [17-19] being enhanced due to electric field confinement in a high-dielectric membrane. Many types of the semiconductor suspended nanostructures, including quantum point contacts [17, 20], quantum dots [19], Hall bars [15], antidot lattices [16, 18] have been implemented. At the same time, suspended semiconductor ring electron interferometers, to our knowledge, have not yet been studied.

The electron-electron interaction is known to play a crucial role in electron phase decoherence [21]. In the suspended 2DEG, new unusual conditions associated with the enhanced inter-electron interaction are realized. Technological difficulties in fabrication of the suspended devices with sizes small enough compared to the phase coherence length are a serious problem that needs to be solved in order to study the electronic interference in the suspended nanostructures. The minimum size of a semiconductor electron interferometer is limited by the size of the depletion region surrounding the hole of the ring rather than by the electron lithography resolution. To provide access to the sacrificial layer which should be removed from under the nanostructure, fabrication of suspended nanostructures requires preliminary deep vertical etching of the heterostructure with a 2DEG, which in turn, enhances the width of the depleted region in comparison with the case of shallow etching. In this case, the contribution of the scattering on edge roughness also increases. Theoretical prediction of the role of all the described features can hardly be made totally convincing. In this connection, the interference phenomena in the suspended 2DEG require experimental investigation.

In the present paper, we report on the fabrication of the suspended semiconductor ring electron interferometer and the study of its magnetoresistance. We observed the Aharonov-Bohm oscillations before as well as after the suspension and studied their temperature suppression.

2. Experimental

The suspended semiconductor ring interferometers were fabricated on the basis of molecular-beam-epitaxy-grown GaAs/AlGaAs heterostructures with a 2DEG. The heterostructures contain 2DEG in a thin GaAs layer at a depth of 90 nm beneath the surface and a 400 nm-thick $Al_{0.8}Ga_{0.2}As$ sacrificial layer. Total thickness of the layers grown above the sacrificial one, i.e. the thickness of the future suspended nanostructures, is 160 nm (see figure 1 (a)). The electron mobility and the 2DEG density are 2×10^6 cm^2/Vs and $(6-7) \times 10^{11}$ cm^{-2} at 4.2 K, respectively. The lateral form of the samples was defined using electron-beam lithography followed by reactive ion etching. The trenches defining the interferometers were etched to a depth of at least 160 nm to provide access of the liquid etchant to the sacrificial layer (see figure 1). The samples were suspended by means of selective wet etching of the sacrificial layer from under the created nanostructures in 1:100 hydrofluoric acid water solution. The created suspended semiconductor ring interferometers have an effective radius 400 – 900 nm. A typical suspended interferometer is shown in figure 2.

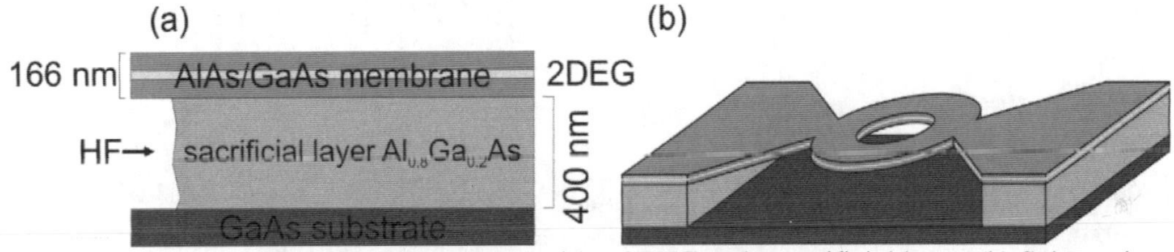

Figure 1. (a) GaAs/AlGaAs heterostructure with a 2DEG and a sacrificial layer. (b) Schematic representation of a suspended ring interferometer.

Figure 2. SEM image of suspended interferometer.

The samples were equipped with Au/Ni/Ge Ohmic contacts to provide the conventional four-terminal magnetoresistance measurements. The measurements were carried out using the lock-in technique in the linear response regime with the alternating current of the magnitude 10 nA and the frequency 7 – 70 Hz at low temperature from 1.6 to 4.2 K. The magnetic field was applied

IOP Conf. Series: Journal of Physics: Conf. Series **964** (2018) 012008 doi:10.1088/1742-6596/964/1/012008

perpendicularly to the 2DEG plane and varied in a range 0 – 0.6 T. The same measurements were performed on the same samples before and after suspension.

At low temperature and zero magnetic field, the rings have low resistance 0.5 – 2 kΩ depending on the illumination and cooling procedure. The value of the resistance (in comparison with $h/2e^2$~12.9 kΩ) apparently indicates that the electron transport through the ring is multimode. The Aharonov-Bohm h/e oscillations are clearly observed in magnetoresistance of interferometers, both before and after suspension (see figure 3 (a – b)). The relative amplitude of the Aharonov-Bohm oscillations $\Delta R/R$ reaches 0.1 – 0.6 %. The fast Fourier transform (FFT) analysis indicates that the oscillations are periodic in the magnetic field with a characteristic period of 6 – 7 mT (see figure 3 (c – d)) that corresponds to the ring effective radius 650 nm. The amplitude of the oscillations demonstrates beating, which can be explained by the interference of several electronic modes.

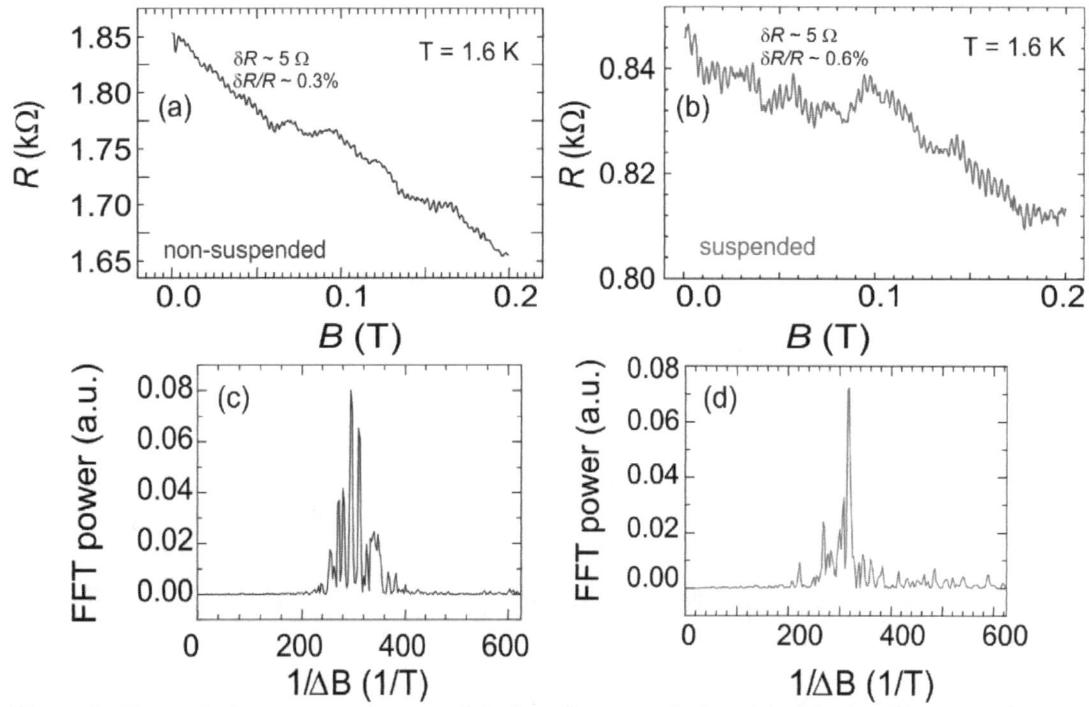

Figure 3. The typical magnetoresistance of the interferometer before (a) and after (b) suspension. Fourier spectra of Aharonov-Bohm oscillations of magnetoresistance before (c) and after (d) suspension.

One can see that absolute amplitude of the Aharonov-Bohm oscillations remains unchanged after the suspension (~ 5 Ω), while the relative amplitude doubled due to a twofold decrease in resistance. Usually the mobility of a 2DEG decreases after suspension [15], which leads to an increase in resistance. The possible reason of the decrease observed in our case is the change in charge state of triangular quantum dots, that are usually formed in the inlet and outlet and affect the electron transport through the interferometer [22]. After suspension of the interferometers, the relative amplitude of the oscillations increases, and the peak in the Fourier spectrum of the oscillations becomes narrower, indicating that the number of propagating modes in the interferometer becomes smaller. Since the absolute amplitude of the Aharonov-Bohm oscillations remains unchanged after the suspension, while the effective radius of the interferometer does not change, as indicated by the peak positions in the Fourier spectrum, it can be concluded that the phase coherence length also remains almost unchanged.

We studied the temperature dependence of the amplitude of the Aharonov-Bohm oscillations in a range from 1.6 to 4.2 K. The amplitude was estimated as the difference in values between two adjacent extrema in a region of the magnetic field where this difference is maximal. The typical dependence for the suspended interferometer is shown in figure 4. Similar behavior is typical for the non-suspended samples. One can see that the amplitude is suppressed by the temperature that confirms the interference origin of the oscillations.

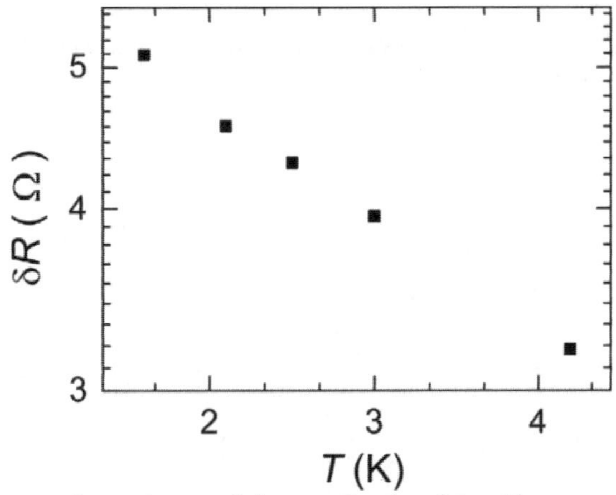

Figure 4. The temperature dependence of the amplitude of the Aharonov-Bohm oscillations.

3. Conclusion

We fabricated a suspended semiconductor ring interferometer with a 2DEG demonstrating oscillations in magnetoresistance, whose temperature dependence and the Fourier spectra are characteristic of the Aharonov-Bohm effect. Relative amplitude of the observed oscillations reaches the large value 0.6%. Relative amplitude of the oscillations increases and the peak in the Fourier spectrum of the oscillations becomes narrower after the suspension, showing that the number of propagating modes in the interferometer becomes smaller, while the phase coherence length remains almost unchanged. The observed temperature suppression of the oscillations indicates their quantum-interference origin.

Acknowledgements
The authors are grateful to Maxim V. Budantsev for help with the experiment. The work is supported by RFBR (Grants Nos. 16-32-00405 mol_a and 16-02-00579) and State Programme (Grant No. 0306-2016-0015).

References

[1] Beenakker C W J, van Houten H 1991 Solid State Physics **44** 1
[2] Timp G, Chang A M, Cunningham J E, Chang T Y, Mankiewich P, Behringer R and Howard R E 1987 Phys. Rev. Lett. **58** 2814
[3] Ford C J B, Thornton T J, Newbury R, Pepper M, Ahmed H, Peacock D C, Ritchie D A, Frost J E F and Jones G A C 1989 Appl. Phys. Lett. **54** 21
[4] Ismail K, Washburn S and Lee K Y 1991 Appl. Phys. Lett. **59** 1998-2000
[5] Casse M, Kvon Z D, Gusev G M, Olshanetskii E B, Litvin L V, Plotnikov A V, Maude D K and Portal J C 2000 Phys. Rev. B **62** 2624-2629
[6] Ekinci K L and Roukes M L 2005 Rev. Sci. Instrum. **76** 061101
[7] Cleland A N and Aldridge J S, Driscoll D C and Gossard A C 2002 Appl. Phys. Lett. **81** 1699
[8] Shevyrin A A, Pogosov A G, Bakarov A K and Shklyaev A A 2016 Phys. Rev. Lett. **117** 017702
[9] Shevyrin A A, Pogosov A G, Budantsev M V, Bakarov A K, Toropov A I, Rodyakina E E and Shklyaev A A 2015 Appl. Phys. Lett. **106** 183110
[10] Shevyrin A A, Pogosov A G, Budantsev M V, Bakarov A K, Toropov A I, Ishutkin S V, Shesterikov E V and Arakcheev A S 2013 Appl. Phys. Lett. **103** 131905
[11] Steele G A, Hüttel A K, Witkamp B, Poot M, Meerwaldt H B, Kouwenhoven L P, van der Zant

H S J 2009 Science **325** 1103
[12] Lassagne B, Tarakanov Yu, Kinaret J, Garcia-Sanchez D, Bachtold A 2009 Science **325** 594
[13] Okazaki Y, Mahboob I, Onomitsu K, Sasaki S and Yamaguchi H 2016 Nature Communications **7** 11132
[14] Yamaguchi H, Okamoto H, Ishihara S and Hirayama Y 2012 Appl. Phys. Lett. **100** 012106
[15] Pogosov A G, Budantsev M V, Zhdanov E Yu, Pokhabov D A, Bakarov A K and Toropov A I 2012 Appl. Phys. Lett. **100** 181902
[16] Zhdanov E Yu, Pogosov A G, Budantsev M V and Pokhabov D A 2013 AIP Conf. Proc. **1566** 211
[17] Shevyrin A A, Pogosov A G, Budantsev M V, Bakarov A K, Toropov A I, Ishutkin S V and Shesterikov E V 2014 Appl. Phys. Lett. **104** 203102
[18] Zhdanov E Yu, Pogosov A G, Budantsev M V, Pokhabov D A and Bakarov A K 2017 Semiconductors **51** 8
[19] Pogosov A G, Budantsev M V, Shevyrin A A, Plotnikov A E, Bakarov A K and Toropov A I 2008 JETP Lett. **87** 150
[20] Okazaki Y, Mahboob I, Onomitsu K, Sasaki S and Yamaguchi H 2013 Appl. Phys. Lett. **103** 192105
[21] Altshuler B L, Aronov A G and Khmelnitsky D E 1982 J. Phys. C **15** 7367-7386
[22] Tkachenko V A, Bykov A A, Baksheev D G, Tkachenko O A, Litvin L V, Latyshev A V, Gavrilova T A, Aseev A L, Estibals O and Portal J C 2003 JETP Lett. **97** 317-330